Introduction to Experimental Nonlinear Dynamics

Nonlinear behavior can be found in virtually every area of science and engineering. Population biology, cardiac fibrillation, laser devices, and aircraft wing flutter are among the highly disparate areas in which nonlinear dynamics and chaos have a role. Largely because of this extensive reach, nonlinear dynamics has become a very active field of study and research.

This book uses an extended case study – an experiment in mechanical vibration – to introduce and explore the subject of nonlinear behavior and chaos. Beginning with a review of basic principles, the text then describes a cart-on-a-track oscillator and shows what happens when it is gradually subjected to greater excitation, thereby encountering the full spectrum of nonlinear behavior, from simple free decay to chaos.

Experimental mechanical vibration is the unifying theme as the narrative evolves from a local, linear, largely analytical foundation toward the rich and often unpredictable world of nonlinearity. Overall, the contents span the range from elementary concepts to sophisticated behavior, using the latest experimental and numerical techniques. Appendices provide further insight with brief descriptions of behavior in an electric circuit analog of the mechanical system and in an aerospace system.

Advanced undergraduate and graduate students, as well as practicing engineers, will find this book a lively, accessible introduction to the complex world of nonlinear dynamics.

Lawrence N. Virgin is a member of the faculty of the School of Engineering and of the Center for Nonlinear and Complex Systems, Duke University.

Introduction to Experimental Nonlinear Dynamics

A Case Study in Mechanical Vibration

Lawrence N. Virgin

 CAMBRIDGE
UNIVERSITY PRESS

PUBLISHED BY THE PRESS SYNDICATE OF THE UNIVERSITY OF CAMBRIDGE
The Pitt Building, Trumpington Street, Cambridge, United Kingdom

CAMBRIDGE UNIVERSITY PRESS
The Edinburgh Building, Cambridge CB2 2RU, UK   http://www.cup.cam.ac.uk
40 West 20th Street, New York, NY 10011-4211, USA   http://www.cup.org
10 Stamford Road, Oakleigh, Melbourne 3166, Australia
Ruiz de Alarcón 13, 28014 Madrid, Spain

First published 2000

Printed in the United States of America

Typeface Times 11/14 pt. and Futura   *System* LATEX 2_ε [TB]

A catalog record for this book is available from the British Library.

Library of Congress Cataloging in Publication Data
Virgin, Lawrence N., 1960–

Introduction to experimental nonlinear dynamics : a case study in
mechanical vibration / Lawrence N. Virgin.

p.   cm.

Includes bibliographical references and index.

ISBN 0-521-66286-9

1. Dynamics.   2. Nonlinear theories.   I. Title.
QA845.V57   2000
003′.85 – dc21   99-26994
                                                            CIP

ISBN 0 521 66286 9 hardback
ISBN 0 521 77931 6 paperback

Contents

Preface

General Comments

• Aims and Scope

Nonlinear dynamics and chaos have been popular areas of research for more than twenty years. One of the most interesting aspects of nonlinear behavior is its ubiquity. Population biology, celestial mechanics, cardiac fibrillation, and aircraft wing flutter are just a few of the highly disparate areas in which nonlinear dynamics and chaos have shed new light. Interdisciplinary research and the cross-fertilization of ideas between different areas has provided much of the impetus. There is hardly a branch of applied science that is not touched by nonlinear dynamics: Witness the range of interdisciplinary books and meetings spread within, and between, research boundaries. However, it is the sheer intrigue provided by nonlinear systems, and chaos in particular, that has caught the imagination of so many people in science and engineering and has led to continued growth in the field.

The aim of this book is to provide a relatively concise pedagogical approach to nonlinear dynamics based on the evolution of an experimental paradigm. In contrast to most other books in this general area, this book will use real data, generated from the laboratory, to introduce, rather than occasionally confirm, the fascinating array of behavior encountered even in a relatively restricted subset of the nonlinear world. In this way, the material will span the gap between abstract, often chaos-driven, ideas and practical engineering reality.

This book uses experimental mechanical vibration as the backbone, or theme, on which the material is developed. We focus attention on this specific branch of engineering to tell a story: one of expanding the horizon from a linear to a nonlinear view, without losing sight of the exigent world of experimentally verifiable behavior. An advantage of this approach is that mechanical vibration, as well as being an important area of research in its own right, has a well-founded history based on linear behavior, and using experimental data as a key component emphasizes the robustness of the described behavior. Thus the narrative of the present book evolves from a local, linear, largely analytical, foundation toward the rich and often unpredictable world of nonlinearity. This book does not attempt to cover the whole range of nonlinear phenomena and experimental techniques. For example, parametric excitation (where a system parameter is time-varying), first-order behavior, and difference equations per se are only covered insofar as they are naturally encountered in the study of forced nonlinear oscillators, and detailed experimental descriptions are kept from interfering with the flow and introduction of new behavior. Rather, this book follows an illustrated, focused, case-study approach. This material can be viewed as a primer, where nonlinear behavior, of increasing severity, is gradually revealed.

• Intended Audience

It is envisioned that the intended audience will reflect the relatively wide scope of the material as well as the focused nature of the context. Specifically the book is aimed at:

Those who have a background in mechanical vibration engineering with the inevitable focus on linear behavior. The material covered by the book provides a strong introduction to nonlinearity based on the types of application with which they are familiar. In this way the book makes an effective supplemental text in advanced courses in mechanical vibration.

Those who work with relatively abstract mathematical or computer-based analyses. This book will provide them with real practical data and its use to illustrate the fascinating details of nonlinear
dynamics in a laboratory context.

Those who conduct experiments but in other, related fields. The relative clarity of the modeling and quality of the data, and the close
correlation between theory and experiment, may provide inspi-
ration in the face of experimental studies of complex systems.

Although the primary intended audience for this book is com-
prised of engineers, I have found the physics community very re-
ceptive to the kinds of macroscale experiments used throughout. In
keeping with any text on vibration theory or dynamics a knowledge
of (linear) ordinary differential equations is quite central. No special
knowledge of experimental techniques is assumed, and a large number of references are included in the bibliography for those wishing
to gain deeper insight.

• Organization of Contents

The first chapters consist of some introductory remarks, a brief review of linear theory, and useful concepts and tools. This provides
an appropriate point of departure for the rest of the book.

The experimental paradigm and its underlying mathematical
model is then described. The specific physical system chosen consists of a low-order, cart-on-a-track oscillator, which closely mimics the behavior of Duffing's equation using relatively unambiguous modeling. It is designed, with pedagogical intent, to provide
relatively high-resolution (low-noise) data, and its configuration
can be conveniently altered to accommodate a variety of types of
behavior.

The chapter on free vibration provides a first taste of dependence
on initial conditions, amplitude-dependent frequencies, etc. This
is followed by the forced case with attention initially focused on
periodic behavior and a first glimpse at how a response typically
loses its stability. Both of these chapters include some approximate
analytical approaches.

Chaos, and ways of measuring and quantifying it, are then introduced. This is followed by a study on the escape problem, that is,

determining under what circumstances the system leaves the confines of its local potential energy well.

Two variations of the basic mechanical theme are then investigated. The first considers a system contrived to exhibit a hardening spring restoring force. This serves to highlight the similarities and differences with the behavior detailed in earlier chapters, and an additional focus on stability is developed. The second provides a new ingredient to the nonlinear mix. An elastic rebound resulting from an impact is introduced to an otherwise linear system. This discrete nonlinearity underlies some interesting new behavior.

Next, consideration is given to the double-well system when subject to a two-frequency excitation. This will introduce motion within a four-dimensional phase space together with the characterization of quasi-periodicity.

The final chapter opens the scope to global behavior, with an emphasis on basins of attraction and manifold interaction.

The appendices contain brief descriptions of systems that augment the material from the earlier chapters. The first confirms the ubiquitous nature of nonlinear behavior by revealing qualitatively similar behavior in an electric circuit analog of the mechanical system. Bottlenecking, an interesting feature of the interaction of local and global behavior, is discussed. Finally, the second appendix gives some flavor to the relationship between simple mechanical models and a real (high-order) engineering system: the acoustically excited, thermally buckled panel, a component highly relevant to an aspect of aircraft design and the problem of sonic fatigue.

A related internet site is located at *http://mems.egr.duke.edu/Nonlinear/Book.html* where many of the experiments described in this book can be viewed in video format.

Acknowledgments

This book would not have been possible without the assistance and support of a number of individuals. Typical of most university environments, much of the hands-on research is carried out by graduate students and often in a cooperative team effort. This is particularly the case with experimental work. In my time at Duke University I have been extremely fortunate to work with a collection of extremely bright and hardworking graduate students, and in an intellectually stimulating atmosphere very conducive to collaborative research. Regarding much of the material on which this book is based, I owe a great deal to two people in particular. First, Jim Gottwald built the original track/cart system (originally as part of a graduate class project) and his skill and enthusiasm carried him through many late nights working on the early stages of this project (which was not even part of his doctoral research). Second, Mike Todd brought his considerable technical abilities to bear and developed not only a thorough understanding of the experimental apparatus (including expertise in programming the LabVIEW software) but also a strong grasp of global issues in nonlinear oscillators.

I also owe a good deal to Kevin Murphy who worked on the hardening spring and the panel systems, Phil Bayly who developed various aspects of the nonlinear dynamics laboratory, Steve Trickey who built the electric circuit described in the appendix, and Jonathan Nichols who worked on the two-frequency excitation. Todd Fielder, Chris Begley, Kara Slade, and Matt Brown also made their experimental contributions to the nonlinear dynamics research group.

I have been very fortunate to be able to work in the pleasant and intellectually vibrant environment of Duke University in North Carolina, with the Center for Nonlinear and Complex Systems (CNCS) providing an effective cross-disciplinary platform. Anyone working in the area of nonlinear science will appreciate the benefits of this kind of interaction.

Many thanks are owed to the Engineering Department of the University of Cambridge (Structures Research Group) in England where much of the text was written. There can be few better places than Cambridge to spend a sabbatical and I am grateful to Chris Calladine and Allan McRobie for making my stay a relaxed and productive one.

I owe a great deal to a number of individuals from whom I've learned. Mike Thompson (University College London) and Earl Dowell (Duke University) have both made important contributions to nonlinear dynamics after establishing their expertise in other areas. The experienced researcher will detect their influence throughout the text. Given the material on which this book is based it would be remiss not to mention the seminal contributions made by Francis Moon (Cornell University) and Philip Holmes (Princeton University). The idea of using simple, discrete, mechanical models, including experiments, to illustrate nonlinear dynamic behavior was inspired by James Croll and Alistair Walker's *Elements of Structural Stability* (McMillan, 1972; lamentably out of print), and their lucid treatment of (static) buckling phenomena. I also thank Henri Gavin of the Civil Engineering Department at Duke for use of his shake table for the experiments described in Chapter 13. I have benefitted from many discussions at seminars and conferences over the years, and I especially appreciate the opportunity to share ideas with Ray Plaut (Virginia Tech), who also diligently reviewed the whole manuscript, Joe Knight (Duke University), Joe Cusumano (Penn State University), and the anonymous reviewers of the manuscript for Cambridge University Press.

Finally, I acknowledge the unswerving support of my wife Lianne and our children – Elliot and Hayley, who help to keep everything in perspective.

Chapter 1

Brief Introductory Remarks

This chapter will set out the motivation and define the context for the material contained in this book. The recurrent themes to be described should help to establish what this book is about (and what it isn't).

1.1 The Physical Context

- **Mechanical Vibration**
  This book uses an aspect of traditional mechanics to provide a focus and solid context for the material development. The field of mechanical vibration provides a rich source of material at the interface between applied mathematics and practical design. Classical vibration is very well developed, especially for linear, low-order systems. Progress is being made in stochastics, control, and nonlinear effects. It is this latter aspect of vibration theory (and its experimental verification) that is pursued in this book.

- **Modeling**
  In this book we are primarily interested in low-order, time-varying, deterministic, nonlinear, experimental, mechanical systems! During the modeling process we make a number of trade-offs between the desire for mathematical simplicity and practical usefulness (and

phenomenological interest). Although the experimental paradigm developed is a discrete mechanical system, the extension to continuous systems becomes clear. Classical methods (e.g., Lagrange's equation) are used to derive governing equations, with the major modeling challenge presented by energy dissipation.

• Duffing's Equation

Throughout we will use Duffing's equation. Why focus so much on one type of system? The development of nonlinear dynamics is framed by intense scrutiny of certain archetypal systems, typically named after the researchers who first studied them: van der Pol, Duffing, Lorenz, Hénon, and Rossler. The nonlinear ordinary differential equation that has come to be known as Duffing's equation has particular importance in engineering, and indeed it was first considered by the German experimentalist Georg Duffing to study the hardening spring effect observed in many mechanical systems.

The specific *double-well* form of Duffing's equation, which provides the backbone of this book, offers compelling pedagogical insights. This is partly based on the fact that it is symmetric (about the origin) in a global sense, a feature nominally present in a variety of practical situations. But since the underlying equilibria are offset, it also subsumes a variety of types of asymmetric behavior and is globally bounded. Under certain conditions, this system also effectively exhibits linear behavior, thus providing the base from which nonlinear behavior evolves. Several basic variations to this mechanical theme will also be developed.

Making a thorough investigation of a specific class of equation in which the nonlinearity blooms from a background of linear behavior offers appealing clarity from an instructional perspective. This is especially useful because the bewildering array of nonlinear behavior can at least be viewed against, and contrasted with, the backdrop of familiar linear behavior, and many of the stability characteristics have their origin in a linear context. The organizational framework provided by dynamical systems theory is of assistance, especially in the classification of instability phenomena:

> Knowledge of the generic or typical properties of dynamical systems and their bifurcations, at the very least, allows one

to organize one's expectations in the analysis of a nonlinear system. It is worthwhile to know what to look for before starting out. (Holmes, 1980)

1.2 Experiments

One of the striking characteristics of the evolution of nonlinear dynamics (and classical mechanics for that matter) as a research field is the intertwined strands of theoretical and applied (experimental) studies, where we can view many recent simulation studies as essentially experimental. Despite the spectacular success of computational studies (witness the impressive computational fluid dynamic studies typically conducted by NASA for a new project), there is still an important role to be played by conducting real experiments in the laboratory. The laboratory setting supplies the stepping stone in going from abstract theoretical ideas to practical solutions for specific engineering problems.

This book is not intended to provide all the tools necessary to enable the reader to conduct his or her own experiments. However, even though the apparatus described in this book is by no means sophisticated, the generation of data and presentation of experimental results serves to give credance to the abundance of material based on purely theoretical (numerical) analyses.

1.3 Stability

Ordinary differential equations (ODEs) have proved to be extremely successful in describing continuous time-varying processes in a huge variety of physical systems. Much of the underlying theory takes advantage of the property of linearity, in which a variety of analytic techniques can be used to study systems that undergo smooth changes in their response. However, in nonlinear systems, behavior can change very suddenly and dramatically. The idea of stability permeates many different facets of our existence ranging from a child learning to ride a bike to the stability of the Universe. We are all familiar with the way the stock market fluctuates and the weather changes, both on various time scales, but within the broad scope of engineering applications we can appreciate the desire to understand, and perhaps be able to predict, among other things how the growing roll motion of a ship can lead to its capsizing; how aircraft wings suddenly start to violently flutter; how earthquakes hit and volcanoes erupt causing Tsunamis and toppling structures; and how heart attacks can strike with no apparent warning.

These are certainly big and wide-ranging problems with often dire consequences, and their complete understanding seems a dim, if not impossible, prospect. But we need to walk before we can run, and in this spirit we use a more local and down-to-earth context to consider exactly how the motion of a mechanical system, oscillating in a regular manner, becomes erratic, or undergoes a qualitative change in behavior. We are specifically interested in the robustness of a given response to finite perturbations and how the response changes under slowly changing conditions:

> Many systems are in a slowly evolving environment, so their coefficients and parameters undergo gradual change. Then if the evolving system is in a steady state of equilibrium, periodic oscillation or chaos, the prediction of any sudden change is of crucial importance. (Thompson and Stewart, 1986)

1.4 Predictability

Another characteristic of this book, and indeed most nonlinear studies, is the overarching context of predictability, which looms to an increasingly large extent during the development of the material. Since the consequences of a loss of stability are so often catastrophic it is clearly useful to be able to predict how a dynamic system evolves. This concern has many different levels: We must consider a variety of issues ranging from fundamental modeling questions (e.g., appropriate constitutive relations for material behavior) to our ability to solve large sets of equations. How is predictability defined? Is it by the average properties at a prescribed time, or will the system state be exactly at a specific location in phase space at a specific time? What is the horizon of predictability? How are these questions different in relation to the space and time scales of different problems? We confine our attempt to find answers to these rather big questions by concentrating on simple experimental mechanics.

The question of predictability is almost a mute point in linear systems. In fact, too strong an adherence to a linear view may be quite dangerous:

> Lulled into a false sense of security by his familiarity with the unique response of a linear system, the busy analyst or experimentalist shouts "Eureka, this is the solution" once a simulation settles onto an equilibrium or steady cycle, without bothering to explore patiently the outcome from different starting conditions. (Thompson and Stewart, 1986)

The lack of predictability is still pertinent even when an *exact* model is available. This fact was well known to Henri Poincaré:

> It may happen that small differences in the initial conditions produce very great ones in the final phenomena. A small error in the former will produce an enormous error in the latter. Prediction becomes impossible. (Poincaré, 1921)

The general approach followed in this book thus seeks to answer the following questions: What happens when a system is gradually subjected to greater (periodic) excitation? How does a system typically lose stability? How predictable is this behavior? These questions are answered with a continual reference to experimental data. Although the material contained in this book follows a quite personal and idiosyncratic narrative, I hope the reader will appreciate the case study context, develop a physical feel for the effects of nonlinearity, and enjoy the vicarious pleasure of observing nonlinear behavior in simple experiments.

Chapter 2

Background: Linear Behavior

2.1 Introduction

In this chapter some elementary concepts in linear oscillations are introduced. This is done for a number of reasons. First, we can properly view linear systems as a subset, or special class, of nonlinear systems with the advantage of access to analytical solutions and anticipated familiarity to the reader. Second, some fundamental theory and notation are established. Finally, the material in this section also provides a somewhat simple and predictable counterpoint to the very much more complicated picture when nonlinearities are added to the ingredients giving us an appropriate point of departure.

There is a large selection of books covering linear vibration theory. This subject is included in most standard texts in physics; for example the relevant chapters of Ref. (Feynman, Leighton, and Sands, 1963) are particularly lucid, and more complete treatments from a practical engineering perspective can be found in Refs. (Thomson, 1981; den Hartog, 1984; Steidel, 1989; Kelly, 1993; Tongue, 1996).

2.2 Simple Harmonic Motion

Consider first the simplest form of oscillator: the spring–mass system (Figure 2.1). If we assume initially that there is no energy dissipation

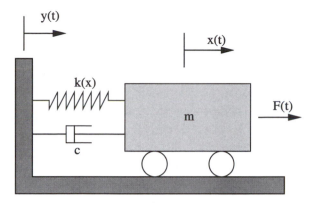

Figure 2.1: A simple spring–mass system.

through damping, or external excitation (i.e., $c = F(t) = y(t) = 0$), and that the spring obeys Hooke's law, applying Newton's laws leads to the second-order, ordinary differential equation

$$m\frac{d^2x}{dt^2} = -kx. \tag{2.1}$$

Throughout this book use will be made of the following notation for time derivatives:

$$\frac{dx}{dt} \equiv \dot{x}, \quad \frac{d^2x}{dt^2} \equiv \ddot{x}, \text{ etc.} \tag{2.2}$$

It is a simple matter to solve Equation (2.1). Imposing some nonzero initial conditions will result in the expected simple harmonic motion (SHM) with a fixed frequency of oscillation. However, before pursuing the form of these solutions further, at this point it is useful to consider an alternative approach to obtaining this equation of motion, based on the concept of energy. This approach has a direct relevance to the choice of mechanical paradigm to be developed in subsequent chapters.

Many physical systems have associated with them an underlying potential, that is, a scalar function that depends only on the starting and finishing points of a path. Consider the scalar potential V and its (gradient) relation to a generalized force $\mathbf{F}$:

$$\mathbf{F} = -\nabla V(\mathbf{x}). \tag{2.3}$$

In the case of unidirectional gravity acting on a point mass, this potential is simply mgy, where y is the vertical distance above some datum, usually taken to be the earth. Hence, using this relation we can write the force acting on the mass as $-mg$. Similarly we can relate the force $-kx$ generated by a Hookean spring to its potential energy $(1/2)kx^2$. Also then, considering the mass m as occupying a single point in space, we can write its kinetic energy as $(1/2)m\dot{x}^2$.

The motion of this type of system, and in general the types of system considered in this book, is conveniently analyzed using Lagrange's equations. They follow from Hamilton's principle (Goldstein, 1980; Marion and Thornton, 1988), which states:

Of all possible paths along which a dynamical system may move from one point to another, within a specified time interval and consistent with any constraints, the actual path followed is that which minimizes the time-integral with respect to the difference between the kinetic and potential energies.

Mathematically, this integral (often called the *action*) can be written as

$$I = \int_{t_1}^{t_2} (T - V)\, dt. \tag{2.4}$$

Introducing the Lagrangian, $\mathcal{L} = T - V$, and specializing the expression for unidirectional motion (in x) leads to

$$I = \int_{t_0}^{t_1} \mathcal{L}(x, \dot{x})\, dt, \tag{2.5}$$

which has a *stationary* value for the correct path of the motion. The calculus of variations can be used to show that the path followed is given by

$$\frac{\partial \mathcal{L}}{\partial x} - \frac{d}{dt}\left(\frac{\partial \mathcal{L}}{\partial \dot{x}}\right) = 0. \tag{2.6}$$

This is Lagrange's equation (Goldstein, 1980), and substituting the expressions for potential and kinetic energy into (2.6) leads to the equation of motion obtained previously. The relationship among Hamilton's principle, Lagrange's equation, and Newtons's laws is fundamental in classical mechanics, and the reader is referred to Goldstein (Goldstein, 1980) for a thorough discussion.

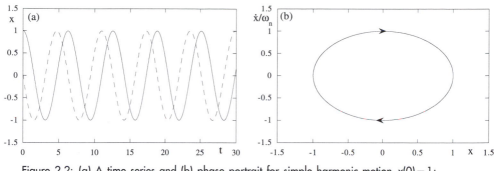

Figure 2.2: (a) A time series and (b) phase portrait for simple harmonic motion. $x(0) = 1$; $\dot{x}(0) = 0$.

In this simple case we can also note the conservation of energy:

$$E = T + V = \frac{1}{2}m\dot{x}^2 + \frac{1}{2}kx^2; \tag{2.7}$$

thus, phase trajectories (ellipses) are carved out in the plane of position versus velocity. This will prove to be a particularly useful representation for nonlinear systems.

It is convenient to introduce the natural frequency $\omega_n = \sqrt{k/m}$ and rewrite the equation of motion (2.1) as

$$\ddot{x} + \omega_n^2 x = 0, \tag{2.8}$$

which has an *exact* solution of the form

$$x(t) = \frac{\dot{x}(0)}{\omega_n} \sin \omega_n t + x(0) \cos \omega_n t, \tag{2.9}$$

where the initial conditions are given by $x(0)$ and $\dot{x}(0)$. The velocity is

$$\dot{x}(t) = \dot{x}(0) \cos \omega_n t - x(0)\omega_n \sin \omega_n t. \tag{2.10}$$

Predictability is self-evident, as simple harmonic motion continues indefinitely. This solution can be obtained using a host of other techniques, including complex algebra or Laplace transforms (Burton, 1994; Ogata, 1998). Figure 2.2 shows a time series (x versus t) and phase portrait (x versus $\dot{x}$), which evolves in a clockwise direction, for a natural frequency of unity (and hence a period of 2π). The velocity is also included as a dashed line in (a) and the quartercycle ($\pi/2$ for the parameters used) phase lag between position and its rate of change is evident.

2.3 Damped Oscillations

2.3.1 Viscous Damping

There is always some dissipation present in mechanical systems: We expect free oscillations to decay. These (nonconservative) forces may take the form of dry (Coulomb) friction, drag (typically proportional to velocity squared), or viscous forces (directly proportional to velocity). This third form of damping is very commonly assumed; one of the motivations for this is that the linearity of the resulting governing equations is not violated by this model. To incorporate damping we use the Rayleigh dissipation function (Meirovitch, 1997) to augment the conservative form of Lagrange's equation (2.6) to get

$$\frac{\partial \mathcal{L}}{\partial x} - \frac{d}{dt}\left(\frac{\partial \mathcal{L}}{\partial \dot{x}}\right) + \frac{\partial P}{\partial \dot{x}} = 0, \tag{2.11}$$

where $P = 1/2c\dot{x}^2$ for linear viscous damping. In this case the nondimensional equation of motion becomes

$$\ddot{x} + 2\zeta \omega_n \dot{x} + \omega_n^2 x = 0, \tag{2.12}$$

where

$$\zeta = \frac{c}{c_c} \tag{2.13}$$

and $c_c = 2m\omega_n$ is a *critical* damping level.

Assuming a simple exponential form for the solution leads to a second-order polynomial (the characteristic equation) whose roots are real or complex depending on the damping ratio. Since complex roots imply oscillatory motion (using Euler identities), this in turn means that the *form* of solutions to Equation (2.12) depend crucially on the damping ratio ζ. For nonzero, positive ζ, three qualitatively different types of behavior are evident:

- Underdamped motion ($\zeta < 1$):
  In this case the solution takes the form

$$x(t) = e^{-\zeta \omega_n t}\left(\frac{\dot{x}(0) + \zeta \omega_n \dot{x}(0)}{\omega_d} \sin \omega_d t + x(0) \cos \omega_d t\right), \tag{2.14}$$

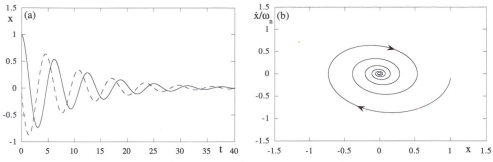

Figure 2.3: (a) A time series and (b) phase plane for underdamped (oscillatory) motion. $x(0) = 1.0; \dot{x}(0) = 0.0; \zeta = 0.1$.

where the damped natural frequency ω_d is given by

$$\omega_d = \omega_n\sqrt{1 - \zeta^2}. \tag{2.15}$$

A typical underdamped response ($\zeta = 0.1$) is shown in Figure 2.3 as a time series and phase portrait. The origin in Figure 2.3(b) indicates the position of asymptotically stable equilibrium. The trajectory gradually spirals down to this rest state; we can imagine a family of trajectories forming a *flow* as time evolves. Since this equilibrium is unique, the whole of the phase space is the attracting set for all initial conditions and disturbances (Abraham and Shaw, 1982). Damping in this range (e.g., $\zeta \approx 0.1$) is quite typical for mechanical and structural systems.

- Critically damped motion ($\zeta = 1$):
  With this specific value of damping, the solution to Equation (2.12) becomes

$$x(t) = [x(0) + (\dot{x}(0) + \omega_n x(0))\,t]\,e^{-\omega_n t} \tag{2.16}$$

  and we note that oscillations cease to exist above this level of damping.
- Overdamped motion ($\zeta > 1$):
  In this case the form of the solution is

$$x(t) = Ae^{(-\zeta + \sqrt{\zeta^2 - 1})\omega_n t} + Be^{(-\zeta - \sqrt{\zeta^2 - 1})\omega_n t}, \tag{2.17}$$

  where

$$A = \frac{\dot{x}(0) + \left(\zeta + \sqrt{\zeta^2 - 1}\right)\omega_n x(0)}{2\omega_n\sqrt{\zeta^2 - 1}} \tag{2.18}$$

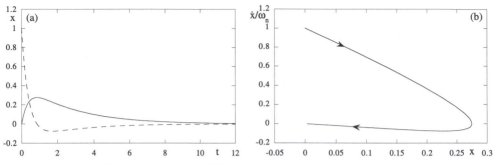

Figure 2.4: (a) A time series and (b) phase plane for overdamped (aperiodic) motion. The initial conditions are: $x(0) = 0.0$; $\dot{x}(0) = 1.0$; $\zeta = 1.5$.

and

$$B = \frac{-\dot{x}(0) - \left(\zeta - \sqrt{\zeta^2 - 1}\right)\omega_n x(0)}{2\omega_n\sqrt{\zeta^2 - 1}}. \tag{2.19}$$

The motion is a generally monotonically decreasing function of time and may take a relatively long time to overcome rather heavy damping forces on the way to equilibrium. A typical case is shown in Figure 2.4.

An observation worth making here is that the mass approaches equilibrium (settles) most quickly in the case with critical damping; this is often a design objective in, for example, the step response of mechanical systems (Burton, 1994). These types of solution will be revisited later from the viewpoint of local attraction to *special* solutions.

2.3.2 Coulomb Damping

It may also happen that energy is dissipated via friction, for example, the sliding of dry (unlubricated) surfaces or rubbing of parts where the magnitude of the damping force, $|F_d|$, is assumed to be independent of speed but its direction always opposes the motion (den Hartog, 1930; Tongue, 1987). In this case it is possible to write two linear ordinary differential equations, each valid for a direction of motion.

It is interesting to note that the amplitude of motion decays linearly in this case and also that motion ceases when a threshold for sticking friction is met. However, because dry friction is essentially a nonlinear mechanism, we shall leave the modeling of Coulomb damping until later.

2.4 Forced Oscillations

Although transient external forces presumably accounted for the nonzero initial conditions to generate the free vibrations of the previous section, it is the general class of harmonic driving force that will be of primary interest in this book. There are a variety of situations where this kind of forced vibration is important, rotating machinery being a good example (Childs, 1993). We will mainly focus on a single harmonic excitation of the form

$$F(t) = F_0 \sin \omega t, \tag{2.20}$$

which can effectively be added to the right-hand side of Equation (2.11). Thus, applying a force directly to the mass in Figure 2.1 results in

$$\ddot{x} + 2\zeta\omega_n\dot{x} + \omega_n^2 x = \frac{F_0}{m} \sin \omega t. \tag{2.21}$$

Equation (2.21) is a nonhomogeneous, linear, ordinary differential equation. Its general solution has both steady-state and transient parts, which combine to give the response:

$$x(t) = \frac{f_0}{k} \frac{\sin(\omega t - \phi)}{\sqrt{[1 - (\omega/\omega_n)^2]^2 + [2\zeta\omega/\omega_n]^2}}$$
$$+ X_1 e^{-\zeta\omega_n t} \sin\left(\sqrt{1 - \zeta^2}\omega_n t + \phi_1\right), \tag{2.22}$$

where $f_0 = F_0/m$ and where trigonometric identities have been used to combine the harmonic terms from Equation (2.14); X_1 and ϕ_1 depend on the initial conditions.

Obviously the second (transient) part of the solution decays with time, leaving the first part as the steady-state oscillation. Some sample responses are shown in Figure 2.5 in which the (lightly damped) system is started from rest at three different forcing frequencies. All the time series plot position as a function of dimensional time (in seconds) rather

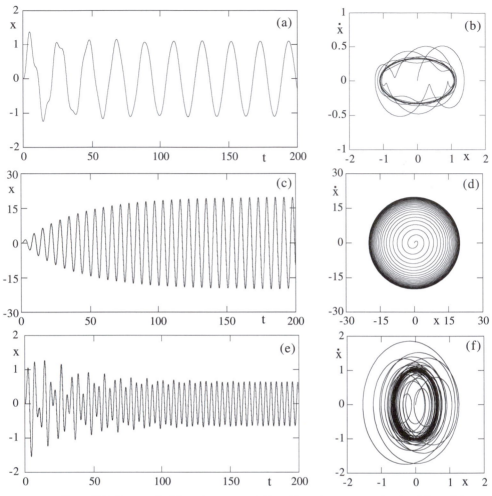

Figure 2.5: Some typical forced responses initiated from the rest state, that is, $x(0) = \dot{x}(0) = 0$, $f_0 = 1$, $\zeta = 0.025$, $\omega_n = 1$. (a) and (b) $\omega = 0.3$; (c) and (d) $\omega = 1.0$; (e) and (f) $\omega = 1.6$.

than nondimensional time (e.g., $\tau = \omega_n t$), and the velocity axis in the phase projections (b, d, and f) are not scaled by frequency. Parts (a) and (b) show that for a forcing frequency (0.3) that is less than the system natural frequency (1.0) the transient is relatively mild compared with the steady-state response and is quickly attracted to the harmonic oscillation. When the forcing frequency equals the natural frequency, as in parts (c) and (d), resonance occurs, that is, there is a significant magnification effect (the denominator in the first term in Equation 2.22 becomes small for $\omega \approx \omega_n$). Note the much larger amplitude of the response. In parts (e) and (f) the forcing frequency is increased to a value of 1.6, and now the transient solution is on the same order of magnitude as the steady-state one, and the steady-state amplitude returns to a lower level. Thus we observe that both parts of the solution depend quite strongly on the frequency ratio. The rate, and hence duration, of the transient decay is primarily a function of the damping. In all these cases the final steady-state motion is independent of the initial conditions (the choice of the origin in Figure 2.5 is representative). This will not necessarily be the case for nonlinear systems, and indeed transients may be repelled by an unstable solution.

It is interesting to summarize how the maximum amplitude of the (steady-state) response ($A = x_{max}k/F_0$) varies with the frequency ratio Ω, where $\Omega = \omega/\omega_n$. By introducing the new forcing amplitude $f_0 = F_0/m$ the normalized amplitude of response can also be written as $A = x_{max}\omega_n^2/f_0$ (Inman, 1994). Clearly the response scales linearly with the forcing amplitude f_0.

Figure 2.6(a) shows a typical amplitude response diagram for four different damping values. The phenomenon of resonance is apparent: A significant amplitude magnification occurs when the forcing frequency is close to the natural frequency (i.e., $\Omega \approx 1$). In fact we see for zero damping a growth to infinite amplitudes. The resonant peak is thus very sensitive to damping ($A_{res} \approx 1/2\zeta$, for light damping), and since many of the nonlinearities of interest are related to larger amplitude motion, we might anticipate interesting behavior in the vicinity of resonantly forced, lightly damped systems. There is also a phase relationship between the input and the output, with the most noticeable feature being a shift (especially for lightly damped systems) at resonance as shown in Figure 2.7. The phase is a measure of how much the output (response) lags behind the input (excitation). Figure 2.7 is based on the mathematical expression

$$\phi = \tan^{-1}\left[\frac{2\zeta\Omega}{1 - \Omega^2}\right], \tag{2.23}$$

obtained during the steady-state solution procedure.

We now briefly describe how this scenario is affected by a transmissible, rather than a direct, excitation of the mass. That is, we assume the system is subject to $y(t) = Y \sin \omega t$, a displacement applied to the

15

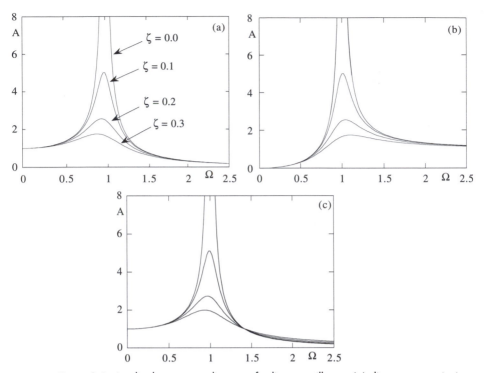

Figure 2.6: Amplitude response diagrams for linear oscillators: (a) direct mass excitation, (b) support motion (relative response), (c) support motion (absolute response). The same damping values are used in (a), (b), and (c).

supporting frame in Figure 2.1. If the position of the mass is measured relative to the moving base (this will be the case for the forced experiments to be described later in the book), then the governing equation of motion takes the form

$$m\ddot{z} + c\dot{z} + kz = mY\omega^2 \sin(\omega t + \phi), \qquad (2.24)$$

where $z(t) = x(t) - y(t)$. Thus, the forcing frequency enters into the forcing amplitude, which is also proportional to the amplitude of the base motion. This is familiar from rotating unbalance (mass eccentricity) in shafts (Childs, 1993). The steady-state amplitude is given by the first part of Equation (2.22) but with an additional ω^2 in the numerator, and the response amplitude is adjusted to take account of the fact that the force now arises via a transmitted base movement ($A = |Z/Y|$, where Z is the amplitude of z). This is shown in Figure 2.6(b).

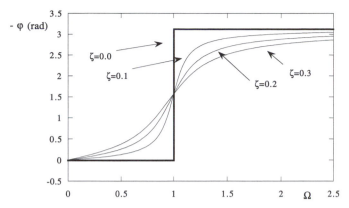

Figure 2.7: The phase–frequency relation corresponding to Figure 2.6(a).

A third form of resonant response (Figure 2.6(c)) is also found for base-excited systems where the *absolute* motion of the mass is measured (here $A = |X/Y|$, where Y and X are the input and response amplitudes respectively (Thomson, 1981)). We note that the resonant condition can be considered as practically the same in all three cases. In this third case the amplitude response includes an interesting independence of damping when the forcing frequency is close to $\Omega = \sqrt{2}$. A brief comparison of these responses shows that in (a) for low excitation frequencies the response is basically the static case where the output (mass position) follows the input (force) nearly in phase, and at high excitation frequencies the mass is practically stationary. In (b) the converse is true. For low frequencies the relative output is zero whereas the static response occurs for high frequencies. Furthermore, we see that in part (c) higher damping levels at high frequencies actually result in a (slightly) larger response. This discussion is relevant to vibration isolation and shock absorbers, where the general goal is to prevent motion being transmitted from structure to support (or vice versa), and to seismic instruments and accelerometers, which seek to *measure* the transmitted motion.

It can also be noted here that although resonance may have considerable practical importance, from a topological standpoint the large-amplitude response is simply a scaling with a smooth growth of behavior. Unlike the nonlinear systems to be described, there are no sudden changes or discontinuities here.

17

2.5 Measurement of Damping

In contrast to the relatively simple matter of the direct measurement of mass or indirect measurement of stiffness, both of which are static in nature, estimating the damping present in a dynamical system is not a simple task, especially since the theoretical basis of energy dissipation is subject to a variety of factors (Levitan, 1960; Holmes, 1979). If a system is acted upon by viscous damping, it is relatively easy to estimate its value based on either free decay tests or resonant effects in forced vibration. Coulomb damping is not so simple to characterize, but reasonable estimates can still be made. Since damping levels tend to be relatively low in the types of mechanics applications of interest here, it can be argued that their approximate modeling suffices for engineering purposes.

2.5.1 Estimating Viscous Damping from a Free Decay

As we have seen, the free decay of a lightly damped spring–mass is typically

$$x = Ae^{-\zeta \omega_n t} \sin(\omega_d t + \phi). \tag{2.25}$$

Taking the logarithm, δ, of the ratio of the magnitude of this curve at times τ_d apart (i.e., successive amplitudes, A_n, A_{n+1}) gives

$$\delta = \ln\left(\frac{A_n}{A_{n+1}}\right), \tag{2.26}$$

which is also related to Equation (2.25) by

$$\delta = \frac{2\pi \zeta}{\sqrt{1 - \zeta^2}}. \tag{2.27}$$

Thus measurement of δ, the *logarithmic decrement* from Equation (2.26), which is relatively easy to make, can be used to estimate ζ. For small values of damping (typical in vibration-type problems) the denominator can be replaced by unity. This idea can be extended to include the decay over a number of cycles, N, which may be a more robust measurement (Tongue, 1987), that is,

$$\delta = \frac{1}{N} \ln\left(\frac{A_0}{A_N}\right). \tag{2.28}$$

2.5.2 Forced Vibration – Half-Power Method

Another popular technique for estimating viscous damping is based on its role in reducing resonant amplitudes. This is called the *half-power method*, and again it relates to lightly damped systems. Assuming that the maximum resonant peak occurs at $\Omega = 1$ (it actually occurs at $\Omega = \sqrt{1 - 2\zeta^2}$ for direct mass excitation) and that it is symmetrical about this value, we can define the resonant peak in terms of a quality factor Q, where

$$Q = \frac{1}{2\zeta}. \tag{2.29}$$

We next determine the frequencies at which the average *power* has dropped to $1/2$ of its resonant value, and denote these by ω_1 and ω_2, where $\Delta\omega = \omega_2 - \omega_1$ is called the bandwidth. Using Equation (2.22) and the above expressions it can be shown that

$$Q \approx \frac{\omega_n}{\Delta\omega}, \tag{2.30}$$

and thus we can determine the damping simply by finding the peak and the frequencies at which it has dropped by a factor of $\sqrt{2}$. This approach has the advantage that since it is based on a ratio, it is not necessary to know the underlying static force acting on the system, which may be difficult to determine in an experiment. This approach is especially popular in acoustics and electric circuits (Beranek, 1954; Carlson and Gisser, 1981).

2.5.3 Estimation of Coulomb Damping

The effect of viscous damping on free decay and its suppression of resonance are relatively straightforward to formulate as shown above. However, this is not the case for Coulomb damping, since the friction force is highly nonlinear. It is still possible, however, to assess the effect of Coulomb damping on both free and forced vibration using a heuristic approach. The modeling of Coulomb damping will be revisited later from a rather theoretical basis for a specific system.

Free Decay

As noted, viscous damping causes an exponential decay of free oscillations. We start by considering the free decay of a simple spring–mass–damper subject to dry friction (characterized by the friction force F_d), which is thus modeled by our familiar second-order ordinary differential equation but valid over one half-cycle (the sign of the damping is always opposite to that of the velocity). Using a work–energy principle, we can obtain an expression for the decay in amplitude per cycle of $(4F_d)/k$, where k is the stiffness coefficient. This linear decay of the amplitude envelope continues until the spring force becomes insufficient to overcome the friction and motion ceases (not necessarily at the origin). The static value of friction for the type of system studied in this book is determined by inclining a track until motion is initiated and then identifying the friction force from the critical angle of tilt. It must be recognized that this estimate is quite approximate since the actual damping process consists of various rubbing and sliding mechanisms.

Forced Vibration

We have also seen how viscous damping tends to reduce the amplitude of oscillation in a simple forced vibration problem, especially in the vicinity of resonance. A simple approximate treatment shows that the maximum amplitude scales inversely with the damping ratio (see Equation 2.29). We have already remarked on the inherently complex nature of dry friction energy dissipation, and the advantages of using a relatively simple mathematical model most appropriate for a specific range of application are clear. For dry friction we can also make some progress toward assessing the role of forced vibration suppression by again appealing to energy balance considerations and developing an equivalent viscous damping model.

Given that the dominant feature of Coulomb damping is to limit the amplitude of resonance, we can again consider the energy lost over a complete cycle of motion. For viscous damping it can be shown that the energy lost per cycle of steady-state harmonic oscillation is given by

$$W_d = \pi c \omega x_{max}^2, \tag{2.31}$$

where c is the damping coefficient and x_{max} is the amplitude of motion at frequency ω. The work done by the Coulomb friction force per cycle of motion is given by

$$W_d = 4F_d x_{max}. \tag{2.32}$$

These expressions can be equated to obtain an *equivalent* viscous damping expression for dry friction:

$$\zeta_{eq} = \frac{2F_d}{x_{max}k\Omega\pi},$$ (2.33)

where the previous definitions for natural frequency and damping ratio have been used. Since we are primarily interested in the resonant effect, we again note that the expression for the frequency response of a viscously damped, *transmissibly forced*, linear oscillator is

$$A = \frac{\Omega^2}{\sqrt{(1 - \Omega^2)^2 + 4\zeta^2\Omega^2}}.$$ (2.34)

Replacing ζ by its equivalent value (Equation 2.33) gives

$$A = \frac{\Omega^2}{\sqrt{(1 - \Omega^2)^2 + \left(\frac{4F_d}{AY\pi}\right)^2}},$$ (2.35)

and, solving for the amplitude, we get

$$A = \sqrt{\frac{\Omega^4 - \left(\frac{4F_d}{Y\pi}\right)^2}{(1 - \Omega^2)^2}},$$ (2.36)

where $A = x_{max}k/Y$. In general, for sliding we will have $F_d = -\mu mg$ for $\dot{x} > 0$ and $F_d = \mu mg$ for $\dot{x} < 0$, where mg is the normal force (weight) and μ is the (material-dependent) kinetic coefficient of friction. We note that in this case the motion does become unlimited at resonance. This expression is valid provided the ratio of the friction force to the magnitude of the applied load remains quite small (typically less than one half). However, in many of the cases considered later, the dry friction force is relatively small and acts in parallel with the viscous damping.

2.6 Superposition

Because this book is largely concerned with nonlinear systems we end this brief introduction by formalizing the concept of linearity. In general, our systems of interest will *not* statisfy the principle of superposition.

Consider the dynamical system:

$$m\ddot{x} + f(x, \dot{x}) = F(t),$$ (2.37)

which governs the time evolution of a point mass, subject to forces that depend on position and velocity, as well as periodic external forces (representative of the types of oscillator studied in this book). We can view this equation as an input–output relation where we seek the output $x(t)$ (position) given a certain input $F(t)$ (force).

Equation (2.37) is said to be linear if, and only if, the following conditions hold:

- For the input $\alpha F(t)$, where α is a constant, the output is $\alpha x(t)$.
- If $x_1(t)$ is the ouput to the input $F_1(t)$, and $x_2(t)$ is the output to the input $F_2(t)$, then $x_1(t) + x_2(t)$ is the output to the input $F_1(t) + F_2(t)$.

In more practical terms we see that for these conditions to hold, we need $x(t)$ and its derivatives to appear to the first power only.

The material introduced in this chapter concerned linear behavior, and the equations of motion (other than the Coulomb damping) clearly satisfied the above conditions. When the principle of superposition holds, a variety of solution techniques can be employed to obtain responses described by relatively simple functions. These exact, closed-form analytical solutions will not, in general, be available for nonlinear systems. However, although linear behavior can be useful in the context of linearization and stability theory, we will generally use numerical integration to solve the governing equations of motion, which will then be compared with experimental data.

Chapter 3

Some Useful Concepts

3.1 Introduction

The previous chapter showed some elementary examples of dynamic behavior largely from a classical (linear) mechanical vibration perspective. Since this book is directed toward nonlinear dynamic behavior, it is important to reexamine these results via the geometric framework provided by the qualitative theory of ordinary differential equations (Guckenheimer and Holmes, 1983; Jordan and Smith, 1977; Krylov and Bogoliubov, 1949; Nayfeh and Balachandran, 1995; Wiggins, 1990). Concepts of stability will begin to take center stage in order to classify the types of qualitative changes in behavior typically encountered in nonlinear systems.

3.2 Phase Space

A fundamental concept in dynamical systems is phase space, in which the observed state of a system traces out a trajectory on a manifold (not necessarily a Euclidian space, but generally smooth). These *noncrossing* paths are solutions to the governing equations, and the velocity vector field is determined from their differentiation (Abraham and Shaw, 1982). We shall see that in nonlinear dynamics it is the behavior of ensembles of trajectories, or flows, and especially their behavior in the vicinity of certain special solutions, that provide considerable geometric insight. Invariant manifolds are objects embedded in the phase space with dimensionality less than the full phase space dimension. They are related to subspaces spanned by eigenvectors and play a special role in organizing dynamic behavior.

We shall focus entirely on deterministic systems, and despite the inevitable noise present in experimental data, we shall consider a trajectory uniquely defined by its state variables. This ties in with the No-Intersection Theorem (Hilborn, 1994), which tells us that two distinct trajectories cannot intersect and that a single trajectory cannot cross itself at a later time. Thus, we note in Figure 2.5 that these trajectories live in a three-dimensional phase space and hence their apparent crossing is really a consequence of the projection through the time axis.

Returning briefly to the damped, unforced, linear oscillator, we can recast the equation in a more useful state variable form:

$$\dot{x} = y,$$
$$\dot{y} = -2\zeta \omega_n y - \omega_n^2 x. \tag{3.1}$$

This system of first-order differential equations not only provides the phase trajectories in a direct form but is also a suitable format for computational treatment (Parker and Chua, 1989).

The state matrix

$$\begin{bmatrix} 0 & 1 \\ -\omega_n^2 & -2\zeta \omega_n \end{bmatrix} \tag{3.2}$$

contains the important characteristics of the dynamics based on the form of the eigenvalues

$$\lambda = -\zeta \omega_n \pm \omega_n \sqrt{\zeta^2 - 1} \tag{3.3}$$

and henceforth are referred to as the characteristic exponents (CEs). We see exactly how the various forms of solution listed earlier depend on the level of damping. We also see how other possibilities include complex CEs with positive real parts. This particular case relates to

$$\ddot{x} - \omega_n^2 x = 0, \tag{3.4}$$

with a solution of the form

$$x(t) = Ce^{\omega_n t} + De^{-\omega_n t}, \tag{3.5}$$

that is, $|x|$ grows without bound as $t \to \infty$.

Returning to the state matrix in Equation (3.2), we can thus classify equilibrium (the origin) according to the nature of the CEs. The lightly damped free response with exponentially decaying harmonic oscillations is associated with a complex pair of CEs whose real parts are small and negative. A family of trajectories evolve in a *spiral* flow to equilibrium. Likewise, the overdamped, monotonically decaying response is associated with real, negative CEs and we classify the equilibrium as a *node*. Given this approach, we can view the unstable response (Equation 3.5) as governing behavior in the vicinity of a *saddle* point with attracting and repelling directions associated with the eigenvectors; initial conditions on the stable manifold will evolve (uncoupled) along that direction toward the equilibrium point. Any inevitable noise or infinitesimal fluctuation will of course excite the unstable component (first term on the right-hand side of Equation 3.5), which soon dominates. We shall see that saddle-type solutions play a crucial role in the global behavior of nonlinear systems.

The idea of phase space is usefully extended to forced oscillations by viewing time as a dependent variable to give

$$\dot{x} = y,$$
$$\dot{y} = -2\zeta\omega_n y - \omega_n^2 x + F\cos\omega t, \tag{3.6}$$
$$\dot{t} = 1.$$

Since we are primarily interested in periodically forced single-degree-of-freedom oscillators, the phase space can be viewed in three dimensions evolving in a toroidal space, that is, $\omega t = \theta$, and hence $\dot{\theta} = \omega$, to give a $\mathbb{R}^2 \times S^1$ phase space (x, y, θ). This three-dimensional flow can be reduced to a two-dimensional map by constructing a surface of section, which will be discussed in the next section.

The phase space also illuminates the concept of energy dissipation from a more global viewpoint via the divergence theorem (Cusumano and Kimble, 1995). For a vector field (flow)

$$\dot{x}_i = f_i(x_j) \tag{3.7}$$

in state variable format, we have

$$div(x_i) = \partial f_1/\partial x_1 + \partial f_2/\partial x_2 + \cdots + \partial f_n/\partial x_n, \qquad (3.8)$$

which will either be zero (without damping) or negative (with damping). We can relate this to the rate of contraction of (small) volume elements:

$$\dot{V}(t)/V(t) = div(x_i), \qquad (3.9)$$

and hence

$$V(t) = V(0)e^{div(x_i)t}. \qquad (3.10)$$

We note at this point that the undamped spring–mass is a Hamiltonian system (from Liousville's theorem (Ott, 1993)), which is thus viewed as an area-preserving system. The systems considered in this book will generally have negative divergence, and for the viscously damped, unforced, linear oscillator we have

$$div(x, y) = -2\zeta \omega_n, \qquad (3.11)$$

which is a constant (equal to the sum of the CEs) and can be considered as the more general context for the estimation of damping based on the logarithmic decrement (Cusumano and Kimble, 1994). Also, for the addition of harmonic forcing we still find the above expression to hold because no contraction or stretching along the time axis occurs. In general the right-hand side is typically the dissipation function, but for linear viscous damping the divergence is a constant (negative) and we can view our systems as totally dissipative. This is in contrast to Hamiltonian systems and their incompressible (or conserved) phase space volumes.

As a final pictorial aid in the concept of phase space consider the situation illustrated in Figure 3.1. This sketch is relevant to the forced oscillator: a three-dimensional phase space (A, R, ϕ), where transients are attracted to an underlying limit cycle (a periodic attractor associated with the first part of the solution in Equation 2.22). This picture also includes a cross section taken at a specific forcing phase (ϕ) where the penetration of the spiraling trajectories contains important dynamic information, and this will be considered in the next section.

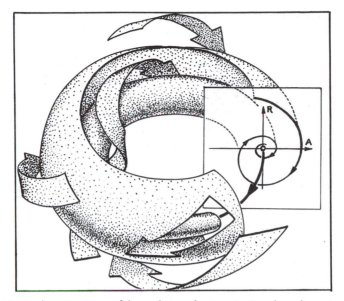

Figure 3.1: A schematic picture of the evolution of trajectories in a three-dimensional phase space and their attraction to a limit cycle. Reproduced with permission from Abraham and Shaw (1982).

3.3 The Poincaré Section

Although plotting amplitude and phase is useful from an engineering perspective, when the stability of periodic attractors is considered (and comparisons with numerical simulation are made) it may be more convenient to construct a stroboscopic Poincaré section, Σ, and plot the position and velocity of the response at a specific forcing phase (Thompson and Stewart, 1986; Moon, 1992). Thus, if the response of the system is given by Equation (2.22), we can set time equal to zero (say), which effectively makes the phase zero. Geometrically we have cut a plane through the three-dimensional phase space (see Figure 3.1). A simpler schematic representation is shown in Figure 3.2 with the first return, or penetration, moving closer to the fixed point x_p than its pre-image (a stable periodic orbit).

A Poincaré section can also be constructed by defining a surface of section in the phase space, for example, by observing the (transverse) penetration of the velocity axis in the unforced oscillator. In some cases

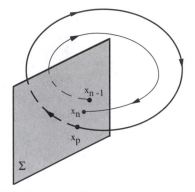

Figure 3.2: Taking a surface of section enables the stability of orbits to be studied as a discrete mapping with a dimension of one less than the original flow.

the time between Poincaré points will not necessarily be constant, but in the periodically forced oscillators of primary interest in this book the forcing phase provides a very convenient Poincaré sectioning trigger.

At this point we use standard trigonometric identities to describe the forced response in the convenient form (the following analysis can also be carried out using complex notation):

$$x_p(t) = a \cos(\omega t) + b \sin(\omega t), \tag{3.12}$$

where the subscript denotes the particular (steady-state) solution, and a and b are related to the amplitude $A = \sqrt{a^2 + b^2}$. Differentiating this equation to get the velocity, we have

$$y_p(t) \equiv \dot{x}_p = -a\omega \sin(\omega t) + b\omega \cos(\omega t), \tag{3.13}$$

and setting $t = 0$ (and hence the forcing phase) we simply get $x = a$ and $y = b\omega$ as the fixed point location. Using the results from the previous chapter, we obtain the exact steady-state solution to this linear system:

$$a = \frac{(1 - \Omega^2)f}{(1 - \Omega^2)^2 + (2\zeta\Omega)^2} \tag{3.14}$$

and

$$b = \frac{2\zeta\Omega f}{(1 - \Omega^2)^2 + (2\zeta\Omega)^2}, \tag{3.15}$$

where the subscript on f has been dropped for convenience.

We can generalize this Poincaré mapping by including the complementary function (the transient solution) in the following way. Consider the system in state variable format:

$$\dot{x} = y,$$
$$\dot{y} = -\omega_n^2 x - 2\zeta \omega_n y + f \cos(\omega t), \tag{3.16}$$

and again including $\dot{\theta} = \omega$, where $\theta = \omega t$, effectively renders this system autonomous, that is, we obtain equations where time does not appear explicitly. Making use of the initial conditions, we can solve for the coefficients in the complementary function (assuming light damping) to get

$$x_h(t) = e^{-\zeta \omega_n t} [c_1 \cos(\omega_d t) + c_2 \sin(\omega_d t)], \tag{3.17}$$

where

$$c_1 = x_0 - a,$$
$$c_2 = \frac{1}{\omega_d}(\zeta \omega_n x_0 + y_0 - \zeta \omega_n a - b\omega), \tag{3.18}$$

and again we have introduced the damped natural frequency $\omega_d = \omega_n \sqrt{1 - \zeta^2}$.

More formally, we construct a cross section at $\theta = 0$ as follows:

$$\Sigma = [(x, y, \theta) \in \mathbb{R}^1 \times \mathbb{R}^1 \times S^1 \,|\, \theta = 0 \in (0, 2\pi)]. \tag{3.19}$$

Thus the Poincaré map is given by

$$P : \Sigma \to \Sigma, \tag{3.20}$$

where both the transient and steady-state components of the general solution are used to obtain

$$\begin{pmatrix} x \\ y \end{pmatrix} \to e^{\frac{-2\pi \zeta \omega_n}{\omega}} \begin{bmatrix} C + \frac{\zeta \omega_n}{\omega_d} S & \frac{1}{\omega_d} S \\ -\frac{\omega_n^2}{\omega_d} S & C - \frac{\zeta \omega_n}{\omega_d} S \end{bmatrix} \begin{pmatrix} x \\ y \end{pmatrix}$$

$$+ e^{\frac{-2\pi \zeta \omega_n}{\omega}} \begin{pmatrix} -aC + \left(-\frac{\zeta \omega_n a}{\omega_d} - \frac{b\omega}{\omega_d}\right) S \\ -b\omega C + \left(\frac{a\omega_n^2}{\omega_d} + \frac{\zeta b\omega_n \omega}{\omega_d}\right) S \end{pmatrix} + \begin{pmatrix} a \\ b\omega \end{pmatrix}, \tag{3.21}$$

where $C \equiv \cos(2\pi \omega_d/\omega)$ and $S \equiv \sin(2\pi \omega_d/\omega)$. This discrete map then has a single fixed point

$$(x, y) = (a, b\omega), \tag{3.22}$$

which was obtained previously by constructing the map on the basis of the steady-state alone. We note that the Poincaré map clearly satisfies the initial conditions. Starting from the initial state of the system, this mapping steps at intervals of the forcing period along the underlying flow. For example, referring back to Figure 2.5, Equation (3.21) describes the progress of the transient, starting at the origin, and stroboscopically samples the solution until settling onto a fixed point. In effect the time axis (coming out of the page in the projections on the right side of Figure (2.5) has been collapsed according to the scheme of Figure 3.1. However, importantly, this complete mapping contains the stability information regarding the fixed point. Although for the linear case this is not really an issue, we shall see later that for nonlinear responses certain exchanges of stability are crucially important.

We also note that this treatment, in the linear context, is suitable for analysis by the Z-transform (Ogata, 1998), the discrete analogue of the Laplace transform. The notion of stability and the relation between these two descriptions will be detailed later.

We have already introduced the concept of characteristic eigenvalues (CEs) determining stability of equilibria in unforced systems. Now we see that it is the eigenvalues of the map (characteristic multipliers – CMs) that determine the stability of cycles based on their trajectory penetration through a defined section of the phase space. The eigenvalues of the Jacobian, $DP(a, \omega b)$, are given by

$$\lambda_{1,2} = e^{-\frac{2\pi \zeta \omega n}{\omega} \pm i \frac{2\pi \omega d}{\omega}}, \tag{3.23}$$

which confirms that the fixed point is asymptotically stable, since both the damping and natural frequency are positive numbers. This is why consideration of discrete maps plays a crucial role in the study of flows (at least in the context of forced oscillators where the reduction from three to two dimensions brings with it considerable advantages). Analogous exact analytic expressions can also be derived for the critically and overdamped cases. An interesting facet of this behavior occurs at resonance: It is instructive to see how transients behave when the system is forced at its natural frequency (i.e., when $\omega = \omega_d$). In this case the map simplifies to

$$\begin{pmatrix} x \\ y \end{pmatrix} \rightarrow e^{-\frac{2\pi \omega n \zeta}{\omega}} \begin{bmatrix} 1 & 0 \\ 0 & 1 \end{bmatrix} \begin{pmatrix} x \\ y \end{pmatrix} + \begin{pmatrix} a\left(1 - e^{\frac{-2\pi \zeta \omega n}{\omega}}\right) \\ \omega b\left(1 - e^{\frac{-2\pi \zeta \omega n}{\omega}}\right) \end{pmatrix}. \tag{3.24}$$

The fixed point is still $(x, y) = (a, \omega b)$ but we now have identical eigenvalues

$$\lambda = e^{-2\pi \zeta \omega_n / \omega}, \tag{3.25}$$

so that there is no spiraling of the Poincaré points as they get mapped onto the fixed point. For example, applying the above mapping to the resonant oscillator (Smith and Smith, 1990)

$$64\ddot{x} + 16\dot{x} + 65x = 64 \cos t, \tag{3.26}$$

with the initial conditions $x(0) = \dot{x}(0) = 0$, leads to the sequence $(0, 0) \rightarrow (0.135, 2.15) \rightarrow (0.197, 3.143) \rightarrow (0.225, 3.598) \rightarrow (0.238, 3.807)$ etc., as the transient is swept (in a straight line) onto the steady state represented (at this forcing phase) by $(0.249, 3.984)$. This is how a Poincaré mapping would evolve for the example shown in Figure 2.5(d), although that example was based on a smaller damping ratio and a very *slightly* different frequency ratio (forced at ω_n rather than ω_d). This situation is somewhat analogous to the transient motion in the case of the critically damped unforced oscillator, with the point attractor at the origin replaced by a periodic attractor defined by the parameters of the problem and the particular section, or phase, at which the trajectory is sampled. In general, however, the Poincaré points will map through spirals since the governing CMs are typically complex.

Despite the fact that this approach tends to hide the usually important engineering aspects of amplitude and phase, it does provide a very convenient means of assessing stability, and, for the types of nonlinear system to be considered later, it provides a powerful tool in the numerical and experimental investigation of periodically excited nonlinear oscillators.

The concept of volumetric contraction of phase space also applies to forced systems, and making use of a Poincaré section we can consider the evolution of areas in a two-dimensional mapping

$$x_{i+1} = G(x_i, y_i),$$
$$y_{i+1} = H(x_i, y_i), \tag{3.27}$$

which results in

$$A_{i+1}/A_i = D = (\partial G/\partial x)(\partial H/\partial y) - (\partial G/\partial y)(\partial H/\partial x) \tag{3.28}$$

for *small* areas A. Here D is the Jacobian determinant, where the eigenvalues are the CMs. We will see later that the product of the CMs is also a constant and can be related back to Equation (3.11). For linear (or linearized) systems this approach is tantamount to the Z-transform analysis popularly used in electric circuit analysis.

In a nonlinear system the analytic expression for the original solution is not in general available, but we can still incorporate these ideas based on approximate solutions and linearization, and they prove to be extremely valuable in the context of numerical solutions and real experimental data.

3.4 Spectral Analysis

A powerful tool in the analysis of dynamic behavior is based on the fundamental idea that any waveform can be represented as the sum of a series of sine and cosine waves at different frequencies, that is, a Fourier series. Thus the spectral analysis of time series, for both periodic and nonperiodic signals, has become a major diagnostic tool across a range of scientific and engineering applications and has played a fundamental role in the development of random vibration analysis (Ewins, 1984; Bendat and Piersol, 1986; Newland, 1984). We will first consider the general case of the Fourier analysis of periodic signals.

Suppose we have a periodic oscillation $x(t)$ satisfying $x(t) = x(t + nT)$, where T is the period of the motion and n is a positive integer. The Fourier series of such a function can be written using complex algebra as

$$x(t) = \sum_{n=-\infty}^{\infty} a_n e^{in\omega_0 t}, \qquad (3.29)$$

where the spectral coefficients, a_n, are the amplitudes of the *discrete* frequency components $n\omega_0$, that is, the frequencies are all exact multiples of the basic frequency defined by $1/T = \omega_0/2\pi$. Equation (3.29) can also be expressed in terms of trigonometric functions but the complex exponential form is more compact. These amplitudes are calculated from

$$a_n = \frac{\omega_0}{2\pi} \int_{-\pi/\omega_0}^{\pi/\omega_0} x(t) e^{-in\omega_0 t} \, dt. \qquad (3.30)$$

The coefficients of the Fourier series can be plotted against frequency to produce a Fourier spectrum (Bergé, Pomeau, and Vidal, 1984). A variation of this is to plot the mean square values of the amplitudes to obtain a power spectrum. From Equation (3.30) we note that the first (constant) term gives the average value of the periodic signal over the duration (provided the integrals are evaluated over integer cycles of motion). In cases where an integer number of periods cannot be guaranteed, the phenomenon of spectral *leakage* is a problem. However, *windowing* is used to mitigate against this and is a well developed area of signal processing (Harris, 1978). In this process the Fourier transform of the windowed, sample data sequence is expressed as the complex convolution of the Fourier transform of the original data with the Fourier transform of the windowing function. A variety of functions have been developed, but in general they consist of functions with low side-lobes. This aspect will be revisited later in the chapter on experimental issues.

It is a straightforward matter to evaluate these expressions for simple harmonic motion or the periodic attractors of the forced linear oscillator. Since the response is effectively a sine wave, we obtain a single spike or delta function in the power spectrum.

This is all very well for periodic signals. However, perhaps the most powerful aspect of frequency analysis is that nonrepetitive signals, even, for example, a single square pulse, can still be analyzed by allowing the frequency to approach zero (i.e., allowing the period to extend to infinity), thus opening the way for the consideration of general nonperiodic signals. This effectively results in a continuous spectral density: the Fourier transform. Mathematically, we write the Fourier *integral*

$$x(t) = \int_{-\infty}^{\infty} a(\omega)e^{i\omega t}\, d\omega \qquad (3.31)$$

and the now continuous, but still complex, amplitude expression

$$a(\omega) = \frac{1}{2\pi} \int_{-\infty}^{\infty} x(t)e^{-i\omega t}\, dt \qquad (3.32)$$

as the Fourier pair, and we observe the frequency content as a power spectrum

$$S(\omega) = |a(\omega)|^2. \qquad (3.33)$$

These expressions are generally based on discrete data and conveniently evaluated numerically. This is what makes this approach such a powerful tool in experimental dynamics.

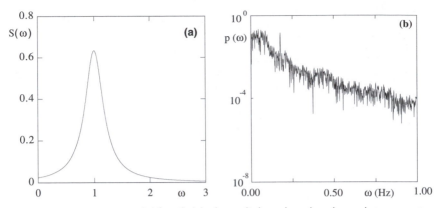

Figure 3.3: Power spectra: (a) free (lightly damped) decay based on the analytic expression for x(t), (b) a typical noisy oscillation based on a numerical evaluation of the FFT.

Consider the example of a transient using the lightly damped, free decay illustrated previously in Figure 2.3. In this case, direct use can be made of the analytic solution for $x(t)$ from Equation (3.32), and evaluating Equation (3.33) (and assuming light damping) leads to

$$S(\omega) = \frac{1}{4\pi^2[4\zeta^2 + (\omega - 1)^2]}, \tag{3.34}$$

which is plotted in Figure 3.3(a). The similarity between this figure and the amplitude response curve for a harmonically *forced* oscillator (Figure 2.6) is evident. The sharpness of the peak is determined by the level of damping, which reminds us of the half-power method introduced earlier.

In the practical (numerical) evaluation of power spectra, advantage can be taken of whether a function is odd or even, and furthermore an efficient algorithm, the Fast Fourier Transform (FFT), allows the rapid evaluation of the Fourier coefficients by repeatedly splitting up the data into shorter sections (Cooley and Tukey, 1965). For example, the result shown in Figure 3.3(a) could also have been obtained numerically by treating the response as a discrete data set (computational efficiency demands 2^n data points). A typical example is shown in Figure 3.3(b) where a noisy time series has produced the rather broadband power spectrum. Here the vertical scale is logarithmic. A crucial point to be appreciated here is that for nonlinear systems this type of power spectrum is often the result of purely deterministic processes. In fact, the signal producing this figure was generated from the numerical solution of a single nonlinear ordinary differential equation, further details of which will be revealed throughout the development of this book.

We also note at this point that the important information provided by a frequency or power spectrum is frequency content. The amplitudes of the various components are not unimportant but they are somewhat sensitive to noise and a variety of scaling options are available for the y axis in these types of plot, for example, power (proportional to amplitude squared), root mean squared (rms) amplitude, or the largest component normalized to one. Of course, their relative peak magnitudes indicate the dominance of certain frequencies. Replotting Figure 3.3(a) with log (or decibel) axes produces the familiar Bode plot from dynamic systems analysis (Burton, 1994; Ogata, 1998). Frequency response occupies such a sufficiently central position in many dynamical studies that a large variety of spectral analysis hardware and software is commercially available.

Before leaving spectral analysis, we also note that white noise (i.e., an uncorrelated random sequence of numbers) would produce a flat power spectrum, which might be considered as the opposite (stochastic) extreme to the (deterministic) delta function of a single harmonic. Needless to say, there are myriad signal processing and numerical subtleties associated with spectral analysis. The problem of leakage was mentioned earlier, but the experimentalist must also be carfeul to have a sufficiently high sampling rate to avoid the problem of *aliasing*, and thus the Nyquist criterion must be satisfied (i.e., only frequencies lower than $f_{Nyq} = f_s/2$, where f_s is the sampling rate, can be resolved). It will be seen that for the types of experiment detailed later, this will not be much of a problem. The specific examples to be developed later in this book are confined to single-degree-of-freedom systems; the reader is referred to Refs. (Newland, 1984; Bergé, Pomeau, and Vidal, 1984) for more information in this area, especially with regard to higher-order systems. This approach will prove to be an extremely useful tool in determining certain properties of nonlinear signals.

3.5 Stability

It might be thought that the example of a system with negative stiffness (Equation 3.5) makes an unlikely candidate for practical interest, but the crucial point here is that even though this kind of linear system (with a repulsive force) is somewhat pathological in isolation (we are presumably always going to be interested in systems that have some initial stability), it is entirely suitable as a description of behavior in the local vicinity of a naturally occurring unstable solution within a *nonlinear* system (Thompson and Stewart, 1986). Furthermore, it is the transition from stable to unstable behavior that is of particular concern. In a nonlinear system, we can expect loss of stability in terms of either perturbations leading to a coexisting solution (global) or a smooth change of parameter(s) leading to a local bifurcation. Bifurcation theory (Golubitsky and Shaeffer, 1985; Troger and Steindl, 1991) provides a general view of this qualitative change in behavior and, in terms of equilibrium configurations, buckling provides a practical context (Thompson and Hunt, 1984). The possibility of sudden, or catastrophic, loss of stability is of particular interest in engineering (Poston and Stewart, 1978; Thompson, 1982; Pippard, 1985).

The preceding concentration on linear behavior is also important in the context of stability theory via linearization, since we will often be concerned with the behavior of *small* perturbations about a solution in order to test its stability.

3.5.1 Stability of Equilibrium

Consider a simple nonlinear oscillator of the form

$$\ddot{x} + \beta \dot{x} + x + x^2 = 0. \tag{3.35}$$

This has two equilibria: $x_e = 0, -1$. To examine their stability we can consider the behavior of small perturbations $x = x_e + \delta$. Placing this in Equation (3.35), we have

$$\ddot{\delta} + \beta \dot{\delta} + x_e + \delta + x_e^2 + 2x_e\delta + \delta^2 = 0. \tag{3.36}$$

If we make the assumption that δ is small then we can neglect δ^2 (and in the general case all higher powers in a Taylor series expansion), and since $x_e + x_e^2$ is identically zero we obtain the linearized equation of motion

$$\ddot{\delta} + \beta \dot{\delta} + \delta(1 + 2x_e) = 0. \tag{3.37}$$

All the preceding linear theory can be used to describe behavior *locally* to these equilibria. Immediately, we see that the $x_e = 0$ position corresponds to decaying (stable) motion, and the $x_e = -1$ position corresponds to exponentially growing (unstable) motion, as described in an earlier part of this chapter. The extent to which these solutions are valid must be carefully considered, but we note that some useful insight can be gained into the full nonlinear behavior of the system by piecing together these essentially local scenarios.

Thus, for our two-dimensional flows (Hirsch and Smale, 1974) we will have

$$\dot{x} = ax + by,$$
$$\dot{y} = cx + dy,$$
$$(3.38)$$

which is the more general case of damped, free vibration governed by Equation (3.1), (i.e., basic rules of mechanics for our class of problem typically ensures the constraint $a = 0$ and $b = 1$). The determinant of this system provides the characteristic exponents (CEs), and the damped forms of free decay can be conveniently related to the discriminant. For stability, we have the requirement that the eigenvalues have negative (or zero) real parts (Pippard, 1985). The movement of these eigenvalues in the complex plane, as a function of a system paramter, gives the root locus (Thomson, 1981).

We also note the relationship between the preceding discussion and the underlying potential energy of a system. For a typical one-dimensional mechanical system with $V = V(x)$, we can write the condition for equilibrium as (Thompson and Hunt, 1984)

$$\frac{dV}{dx} = 0. \qquad (3.39)$$

In general we can expand the potential function as a Taylor series about x_e and can assume, without loss of generality, that the equilibrium point is at the origin:

$$V(x) = V_e + x\left(\frac{dV}{dx}\right)_e + \frac{x^2}{2!}\left(\frac{d^2V}{dx^2}\right)_e + \frac{x^3}{3!}\left(\frac{d^3V}{dx^3}\right)_e + \cdots, \qquad (3.40)$$

We can take the arbitrary constant V_e as 0 at $x = 0$ and since the first derivative is zero from Equation (3.39), we have

$$\frac{1}{2}x^2\left(\frac{d^2V}{dx^2}\right)_e \qquad (3.41)$$

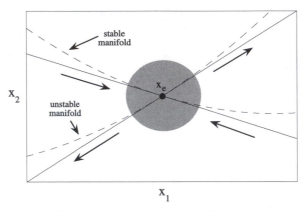

Figure 3.4: A nonlinear flow can be approximated using linearization about an equilibrium point.

as the dominant leading term. We can state the following theorem due to Lagrange (Thompson, 1982):

An equilibrium state at which the total potential energy is an isolated local minimum is necessarily stable.

Thus we can examine the sign of this quadratic form for the purposes of (static) stability. If expression (3.41) is positive (negative) then the equilibrium is stable (unstable). Classic buckling analysis in structural mechanics investigates the underlying potential energy function associated with an equilibrium configuration experiencing a critical point. For example, a structure buckles when the local potential energy function ceases to exhibit a local minimum, (i.e., when expression (3.41) equals zero (Thompson and Hunt, 1984)). The addition of energy dissipation in the form of damping renders these types of local minima asymptotically stable, a stronger form of stability. This type of linearization can be viewed geometrically for two-dimensional flows as in Figure 3.4, where x_1 and x_2 are state variables, say, two generalized coordinates or position and velocity. Here, we can focus attention on the analytically simpler local manifolds and their influence on trajectories in the vicinity of an equilibrium (or fixed) point. This figure shows a saddle point (i.e., an equilibrium that does *not* satisfy the conditions for stability based on Lagrange's theorem). The stable and unstable manifolds are also referred to as the inset and outset of the fixed point. We will see that this type

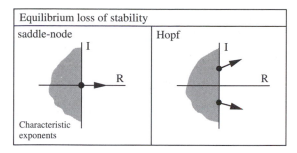

Figure 3.5: Generic routes to instability for a two-dimensional flow under the action of a single control.

of essentially local analysis has a powerful application when applied to unstable periodic orbits.

A useful classification of instability phenomena (i.e., local growth of motion) for these types of systems is based on how the eigenvalues (CEs) exit the negative real half-plane (Pippard, 1985). Figure 3.5 shows the two generic routes when a single (control) parameter of the system is changed. The first mode of instability is classified as a *saddle-node* bifurcation where one of the two CEs becomes positive and real. This phenomenon has also been termed a *fold* since the control and a state variable are often quadratically related, as well as *divergence*, since equilibrium no longer exists. For example, this type of behavior would occur when the spring stiffness drops to zero and thus is highly relevant to the buckling of engineering structures (e.g., the snap-through of a shallow elastic arch (Thompson and Hunt, 1984; Bazant and Cedolin, 1991)). The second generic route to instability occurs when the real part of a complex conjugate pair of CEs becomes positive. This is known as a *Hopf* bifurcation and would occur as the damping present in a system becomes negative. Although this is a commonly occurring situation in certain flow-induced problems (e.g., flutter in aeroelasticity (Dowell, 1975; Holmes, 1977)), it will not typically be encountered for the types of mechanical systems considered in this book.

3.5.2 Stability of Cycles

Nonlinear oscillators may possess a variety of stable and unstable periodic orbits. Their (local) stability can be investigated using Floquet theory based on the general solutions to linear ordinary differential equations with periodically varying coefficients (Lyapunov, 1947; Jordan and Smith, 1977; Nayfeh and Mook, 1978). However, similar results can be obtained using a numerically equivalent approach, which has the dual advantages of not being approximate and of being applicable to experimental data (Murphy et al., 1994). The stability of periodic oscillations can be treated in an analogous way to the stability of equilibria via the Poincaré sampling introduced earlier (Hénon, 1982). For the single oscillator-type systems of primary interest, we will have a three-dimensional flow stroboscopically sampled to produce a two-dimensional mapping, and we shall see that it is now the mapping eigenvalues that determine the stability of the underlying cycle.

Suppose we have a periodic solution $x(t)$ (e.g., the typical response of a harmonically forced, damped oscillator) satisfying

$$x(t) = x(t + T). \tag{3.42}$$

Following Refs. (Hartman, 1964; Guckenheimer and Holmes, 1983), we again appeal to the concept of a linear description of behavior in a local vicinity of this periodic solution from the variational equation

$$\dot{\xi} = Df(x(t))\xi, \tag{3.43}$$

where $Df(x(t))$ is a 2×2, T-periodic matrix evaluated at the known periodic orbit. Any fundamental solution matrix can then be written in the form

$$X(t) = Z(t)e^{tR}, \tag{3.44}$$

where $Z(t) = Z(t + T)$, and X, T, and R are 2×2 matrices. Choosing $X(0) = Z(0) = I$ leads to

$$X(T) = Z(T)e^{TR} = Z(0)e^{TR} = e^{TR}, \tag{3.45}$$

such that the eigenvalues of the constant matrix e^{TR} govern the behavior of solutions in the neighborhood of the periodic orbit. This in turn is directly related to the Poincaré map, which, although available analytically for the linear oscillator, will generally be obtained numerically without much difficulty (Gear, 1971). The above approach can be generalized to include the stability of subharmonic solutions and higher-order systems.

The eigenvalues of the Jacobian matrix (Equation 3.45) are called the Floquet, or characteristic, multipliers (CMs) and take on a discrete mapping analogy to the characteristic exponents that govern the stability of equilibria for continuous flows. Now the condition for stability is that the *magnitudes* of the CMs must be less than unity for stable periodic behavior, with the unit circle replacing the imaginary axis as the stability transition. The relationship between CEs and CMs is based on the exponential map and involves a simple transformation incorporating polar coordinates (Abraham and Shaw, 1982), that is,

$$CM = e^{CE}, \tag{3.46}$$

where the axes of the CE plane have been shifted and wrapped around the unit circle in the CM plane. Thus, stability transitions based on a horizontal progress through the CE plane get mapped into rays emanating from the origin in the CM plane, and the *time unit* for this CE is the time between Poincaré sections (typically the forcing period). Control engineers are very comfortable shifting between the continuous s-plane and the discrete Z-plane in their study of linear control systems where the avoidance of instability is a major issue (Ogata, 1998). A nice pictorial view of this transformation can be found in Ref. (Abraham and Shaw, 1982). Later, we will generalize this to include the concept of Lyapunov exponents (LEs). Note that the divergence theorem tells us that the sum of the CEs (and thus the product of the CMs) in Equation (3.46) is a (negative) constant, that is, on average volumes contract in the (continuous) evolution of the flow, and areas contract in the (discrete) evolution of the map (based on the Poincaré section).

Focusing attention on the types of forced oscillator of most interest, we consider the typical two-dimensional maps of the form

$$\begin{aligned} x_{i+1} &= ax_i + by_i, \\ y_{i+1} &= cx_i + dy_i \end{aligned} \tag{3.47}$$

(i.e., the linearized form of Equation 3.27 and the discrete analogue of Equation 3.38). Again it is the eigenvalues and eigenvectors that contain the pertinent stability information (Abraham and Shaw, 1982). A periodic orbit, or limit cycle, with none of its CMs directly on the unit circle is termed hyperbolic.

The generic routes to instability of a periodic orbit under the action of a single control are shown in Figure 3.6. Later in this book we will focus attention on the manifestation of these first two types of instability in the context of harmonically forced nonlinear oscillators. We simply note that the first instability mechanism is often associated with a finite

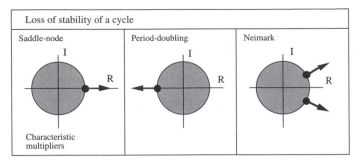

Figure 3.6: Generic routes to instability of a cycle for a system under the action of a single control.

jump in response with a periodic solution being replaced by another (remote) solution. The second instability mechanism represents a bifurcation from period-one to period-two behavior. This event may also signal the start of a complex sequence of events leading to chaos. Again, the physical constraints of mechanical systems will result in these first two forms of instability being more commonly encountered, although the Neiwmark bifurcation plays an important role in systems exhibiting quasi-periodicity.

Before leaving this discussion of stability, a couple of additional points should be mentioned. This brief outline of stability has focused on generic routes. There are other instabilities associated with both pathological cases, for example, symmetry-breaking (pitchfork), and a variety of bifurcational behavior can occur in systems with many controls and higher-order phase spaces (Poston and Stewart, 1978; Nayfeh and Balachandran, 1995). Also, since the stability of equilibria and cycles is reflected in the behavior of small superimposed perturbations, it may be possible to predict an incipient instability from nondestructive testing (Virgin, 1986). It has also been shown that some of these instabilities may have noisy precursors that appear in power spectra before the event occurs (Wiesenfeld, 1985).

3.6 Numerical Issues

Despite the readily available closed-form solutions for linear systems, and the historical importance of approximate analyical methods for weakly nonlinear systems, we generally turn to computational techniques to explore the breadth of nonlinear behavior from a theoretical standpoint. Because the main emphasis in this book is on experimental behavior, only relatively occasional use is made of simulation results, and generally just for verification purposes. Hence, only a brief introduction is given and the references listed can be consulted for details.

A number of numerical techniques are used in the process of data analysis and signal processing (e.g., filtering out the effects of noise (Beckwith, Marangoni, and Lienhard, 1993)). In some of the later results a finite difference scheme will be used to extract velocity information from a time series. However, there are problems associated with the amplification of noise, and in general, time-lag embedding will obviate the need to extract velocity directly (see Chapter 6).

Several general computational tools will also be used later in this book, including the FFT (see Section 3.4), which has come to be used as a key diagnostic tool throughout dynamics and vibration and is subject to a variety of numerical issues including aliasing and leakage (Ewins, 1984; Newland, 1984). Some of the fixed-point characterization to be described in Chapters 9 and 11 makes extensive use of the method of least squares (Beckwith, Marangoni, and Lienhard, 1993). Newton–Raphson and the shooting method are also called upon to assist in the solution process (Parker and Chua, 1989). Other specialist tools developed to shed light specifically on chaos will be discussed in Chapter 9.

However, it is simulation, based on rapidly expanding computational power, that has underpinned the rapid growth of interest in nonlinear dynamics. The numerical solution of nonlinear ordinary differential equations (initial value problems) occupies a rather central place in applied mathematics and a vast assortment of techniques are available (Gear, 1971). In general, a fourth-order Runge–Kutta algorithm is used in this book unless stated otherwise. This popular method combines speed and accuracy and is available in a variety of formats (Press et al., 1992). The use of a constant time-step, although not necessarily the best approach in terms of optimizing the procedure, is generally adopted here, since the extraction of Poincaré points can then be achieved quite easily. We note at this point that if an arbitrary surface of section is chosen, for example one defined by the trajectory penetration of a plane within a phase space in an autonomous system, the method of inverse interpolation can be conveniently used (Hénon, 1982). The (numerical) initial condition plots to be shown in Chapter 14 and Appendix A are based on direct numerical integration, although the method of cell-to-cell mapping is an efficient alternative (Hsu, 1987).

An important recent development involves the use of continuation methods, in which solution paths are followed (Kaas-Petersen, 1987; Parker and Chua, 1989; Seydel, 1991). These methods have the advantage that unstable solutions can also be obtained (we shall see their global importance later); such solutions are not ordinarily attainable using direct numerical integration. The stability of solutions can then be studied in a straightforward manner.

Although most of the numerical results to be presented were based on conventional programming (generally FORTRAN and C), increasing use is being made of software libraries (e.g., *IMSL* (Math/Library, 1989) and *Matlab* including *Simulink* (Matlab, 1989)); algebraic manipulation packages (e.g., *Mathematica* (Wolfram, 1996) and *Maple* (Robertson, 1996)); general dynamics programs (e.g., *dstool* (Guckenheimer and Worfolk, 1993) and *Dynamics* (Nusse and Yorke, 1994)); and specialist dynamics programs (e.g., *AUTO* (Doedel, 1986) and *Bifpack* (Seydel, 1994)). The data acquistion software used throughout (LabVIEW) (Chugani, Samant, and Cerna, 1998) also contains a wealth of computational tools. One of the particular advantages of using these packages is that sophisticated graphics are routinely made use of – a feature that we shall find absolutely crucial as we delve into the world of nonlinear dynamics.

3.7 Extensions to Higher-Order Systems

Although the preceding discussion and the material contained in this book in general are based on low-order systems with just a single mechanical degree of freedom, many of the ideas can be extended to cover higher-order systems. Often this will involve a not unexpected increase in algebra, together with a partial loss of insight owing to sheer geometric complexity. For linear systems this extension to higher-order systems is relatively routine, but we will observe the vastly increased range of dynamic possibilities in going from a two-dimensional to a three-dimensional flow for nonlinear systems. It is certainly true to say that there remain many open issues concerning the dynamics of higher-order systems with a natural progression culminating in continuous systems governed by partial differential equations (Holmes, 1990).

However, it is always important to understand the simple systems first, especially for experimental systems that present a multitude of additional complications and may obscure pedagogical intent.

Furthermore, there are many instances in higher-order systems where much of the significant action takes place within a subspace of the original phase space, ranging from modal analysis in dynamic testing (Ewins, 1984) to the center manifold theory in differential equations (Carr, 1981; Wiggins, 1990). There are also a variety of techniques based on the desire to somehow advantageously project behavior onto a lower dimensional space (e.g., the elimination of passive coordinates (Thompson and Hunt, 1984), Galerkin's method (Meirovitch, 1997), normal forms (Nayfeh and Balachandran, 1995), and Karhunen–Loeve decomposition (Bayly and Virgin, 1993a)). For very high-order systems it is likely that statistics must play a greater role (e.g., a probability density function may be a useful means of extracting global information).

Thus, the emphasis in the rest of this book will be firmly based on low-order systems where the complexity is manifest in temporal evolution. Linear theory works remarkably well in many applications, and the extension to higher-order (continuous) analysis is mostly just a matter of scale. We shall see that nonlinearity presents considerable intrigue for low-order (experimentally verifiable) systems, and this, in turn, opens the entrance doors to temporal and spatial complexity in high-order systems. However, spatially extended systems are beyond the scope of the present book.

Chapter 4

The Paradigm

4.1 Introduction

A considerable amount of research has been conducted on the behavior of nonlinear dynamical systems. Much of this work has been based on numerical integration. Without the luxury of closed-form analytic solutions, coupled with the development of high-speed processors, algorithms, and graphics capabilities for digital computation, it is not surprising that numerical simulation has come to play a central role in many studies (Parker and Chua, 1989).

However, some of the earliest work in nonlinear dynamics was strongly based on experimental observation. Two of the classic early nonlinear oscillators were the van der Pol and Duffing equations, both named after the researchers who wanted to understand their range of behavior and relevance to a practical physical problem: radio oscillations in the former case (van der Pol, 1934) and nonlinear springs in the latter (Duffing, 1918). These two equations have become a proving ground for a host of approximate analytical methods and continue as prime exhibitors of chaos. It is the latter with its emphasis on nonlinear stiffness that will form the kernel of the material on which this book is based.

4.2 Some Basic Forms of Nonlinearity

In the context of experimental mechanics, Duffing's equation has formed the basis for a seminal illustration of chaos (Holmes, 1979; Moon and Holmes, 1979; Holmes and Moon, 1983), and much of the ensuing material is drawn from a relatively simple experimental mimicry of Duffing's equation (Gottwald, Virgin, and Dowell, 1992) and some of its close relatives (Gottwald, Virgin, and Dowell, 1995; Todd and Virgin, 1997a; Todd and Virgin, 1997b; Murphy et al., 1994). Duffing's equation consists of a single-degree-of-freedom oscillator, with a harmonic driving term, and in some sense is the simplest example drawn from a general class of nonlinear oscillators (Poston and Stewart, 1978) given by

$$X'' + 2\zeta X' + R(X) = F \sin(\Omega \tau + \phi), \qquad (4.1)$$

where $X = X(\tau)$. Typically we will be interested in restoring forces that can be approximated as simple polynomials:

$$R(X) = AX + BX^2 + CX^3, \qquad (4.2)$$

with a corresponding potential energy function:

$$V(X) = \int R(X)\, dX = \frac{A}{2}X^2 + \frac{B}{3}X^3 + \frac{C}{4}X^4 + C_0. \qquad (4.3)$$

The signs of the coefficients allow considerable flexibility, and many of these forms occur quite naturally in the context of structural mechanics.

The aforementioned study (Moon and Holmes, 1979; Holmes and Moon, 1983) was based on a magnetically buckled thin elastic beam with an underlying symmetric "double-well" potential energy function, which occurs in Equation (4.2) if $A < 0$, $B = 0$, and $C > 0$, and hence the above mathematical model can provide an effective single-mode approximation to the dynamics of a continuous system.

Figure 4.1 shows the resulting restoring force and associated potential energy expressions for a number of nonlinear forms. Hooke's law is thus given by $A = 1$, $B = C = 0$ with a constant stiffness and parabolic potential energy curve (the continuous line in Figure 4.1). The arbitrary constant (C_0) in the potential energy expressions is taken as zero, and

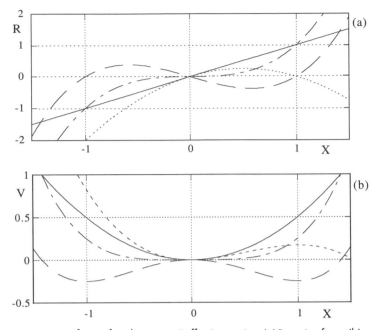

Figure 4.1: Various forms of nonlinearity in Duffing's equation. (a) Restoring force; (b) potential energy.

we shall ignore the damping and external drive force for the moment:

$$X'' + X = 0. \tag{4.4}$$

The double-well Duffing equation ($A = -1$, $B = 0$, $C = 1$; the long-dashed line in Figure 4.1) will form much of the basis for further development in this book:

$$X'' - X + X^3 = 0. \tag{4.5}$$

We see that for small deviations from either stable equilibrium (at $X = \pm 1$) the (horizontally shifted) linear oscillator provides an adequate approximation. Another interesting case is the asymmetric oscillator where we might have a quadratic stiffness term (B) as the only nonlinearity. This can serve as an approximation to moderately large excursions from a single well within the double-well potential (and shown as the short dashed line in Figure 4.1) (Virgin, 1986):

$$X'' + X - X^2 = 0. \tag{4.6}$$

Also included here is the rather pathological case of a purely cubic oscillator (Ueda, 1980) (and shown as a dot–dash line in Figure 4.1):

$$X'' + X^3 = 0. \tag{4.7}$$

We note that the pendulum equation $(A = 1, B = 0, C = -1)$, which has been used to model the restoring force in ship rolling (Virgin, 1987) and is the subject of many studies in nonlinear dynamics, is not considered in this book (Baker and Gollub, 1996). Finally, we can envision an impact barrier located at some point on the parabolic (linear oscillator) potential:

$$X'' + X = 0, \quad X > X_c. \tag{4.8}$$

In this case X is restricted to values greater than X_c, that is, the system experiences a sudden velocity reversal at impact, which can often be modeled using a coefficient of restitution $\dot{X}^+ = -e\dot{X}^-$.

These forms provide a representative sample of nonlinear stiffnesses. It is a simple matter to consider the phase trajectories in these cases for conservative systems. Linearization can also be envisioned from Figure 4.1(a) by fitting a straight line through an appropriate equilibrium $(R = 0)$.

4.3 Nonlinear Structural Dynamics

These nonlinear stiffness functions occur quite naturally in the context of large-amplitude oscillations of thin beams (Krylov and Bogoliubov, 1949; Tseng and Dugundji, 1971; Bazant and Cedolin, 1991). This also provides an example of how continuous (distributed parameter) systems can often be analyzed using just a single (lumped parameter) mode. The following section gives a brief introduction to some simple modeling of thin elastic beams and how they provide a practical context for nonlinear dynamics in general, and Duffing's equation in particular.

Systems with both spatial coordinates and time as independent variables are often formulated in terms of partial differential equations (e.g., the classic wave equation). We will restrict this brief discussion to essentially one-dimensional (beam-type) systems where time (t) and distance along the beam (x) are the two independent variables. In this section we use the symbol x as a coordinate in keeping with standard structural

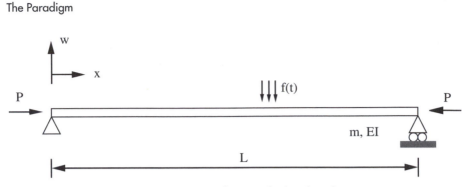

Figure 4.2: A schematic of a thin elastic beam.

mechanics convention (but it should not be confused with the general dependent (position) variable used in the rest of this book). Lateral free oscillations will be considered, although because the primary nonlinearity is a geometrically large deflection we may easily extend these ideas to the forced case. An applied axial load can also be used to promote nonlinear stiffness effects, and indeed adjustment of this parameter is a convenient method of changing the potential energy function along the lines of the previous section. It will be seen that often certain assumptions can be made such that the x-dependence can be effectively eliminated, leaving a Duffing-type ordinary differential equation to model the behavior (i.e., yielding a lumped parameter model (Moon and Holmes, 1979)).

Consider the continuous beam (or strut) shown in Figure 4.2 with mass per unit length m and constant flexural rigidity EI, subject to an axial load P and lateral excitation $f(t)$. The length is L, the coordinate along the beam is x, and the lateral (transverse) deflection is $w(x, t)$. For simplicity, the boundary conditions are assumed to be simply supported (pinned), and linear viscous damping c is assumed to be present. From equilibrium of forces acting on a differential segment of the beam, d'Alembert's principle, and addition of damping forces, we can obtain the governing equation of motion for free vibration (with $f(t) = 0$):

$$m\frac{\partial^2 w}{\partial t^2} + c\frac{\partial w}{\partial t} + EI\frac{\partial^4 w}{\partial x^4} + P\frac{\partial^2 w}{\partial x^2} = 0. \tag{4.9}$$

We initially assume that $P = 0$ and $c = 0$ and that the motion consists of a constant shape $\phi(x)$ that varies with $Y(t)$ such that

$$w(x, t) = \phi(x)Y(t). \tag{4.10}$$

Substituting this into Equation (4.9) and separating variables leads to two *ordinary* differential equations (one in space, which can be solved using the appropriate boundary conditions and describes the mode shapes – half sine waves in this case; the other in time, which can be solved with the appropriate initial conditions and contains the frequency information – simple harmonic motion). It is not uncommon for the vibration of mechanical systems to be dominated by the lowest mode, that is, the mode with the lowest frequency.

However, at this point we focus attention on the underlying statics and make use of an energy approach to gain insight into how beam-type systems are related to the cubic nonlinearity in Duffing's equation. Returning to Figure 4.2, we now ignore the dynamic effects and assume that the axial load causes an end-shortening of ξ. It can be shown that the curvature ψ is related to the lateral deflection via (Bazant and Cedolin, 1991; Thompson, 1982)

$$\psi = w''(1 - w'^2)^{-1/2}, \tag{4.11}$$

where the prime denotes differentiation with respect to the arc length x. We note that a very similar expression (equivalent in its first couple of terms in a series expansion) can be used for curvature based on a horizontal coordinate. It can also be shown (based on inextensional beam theory) that the total strain energy stored in bending is

$$U = \frac{1}{2}EI \int_0^L \psi^2 \, dx. \tag{4.12}$$

Placing ψ in the above expression and expanding, we obtain

$$U = \frac{1}{2}EI \int_0^L (w''^2 + w''^2 w'^2 + \cdots) \, dx. \tag{4.13}$$

Similarly, we can relate the end-shortening to the lateral deflection:

$$\xi = \int_0^L (1 - w')^{1/2} \, dx, \tag{4.14}$$

which in turn can be related to the potential energy of the load:

$$V_P = -\frac{1}{2}P \int_0^L \left(w'^2 + \frac{1}{4}w'^4 + \cdots\right) dx. \tag{4.15}$$

Given a form for w, these two expressions (4.13 and 4.15) can then be added to the appropriate kinetic energy expression as an alternate route to the governing equation of motion via Lagrange's equation (Thompson, 1982).

In the absence of axial load, the potential energy consists of just the strain energy in bending. For small-amplitude vibration, we basically have the situation encountered in Chapter 2, that is, a unique equilibrium configuration with harmonic oscillations about the straight position. For large-amplitude motion, depending on the nature of the constraint at the ends of the beam, a tension effect may be induced, which leads to a hardening spring nonlinearity (Tseng and Dugundji, 1971). However, it is the presence of the axial load that allows the full spectrum of non-linear features and provides the compelling analogy with the twin-well Duffing system.

A classical mechanics result is that the elastic buckling load of a simply supported strut (like the one in Figure 4.2) is given by the Euler load (Croll and Walker, 1972; Bazant and Cedolin, 1991):

$$P^c = EI\left(\frac{\pi}{L}\right)^2. \tag{4.16}$$

Provided the axial load remains less than this value, the trivial equilibrium state is stable. When the axial load is identical to this value the stiffness effectively drops to zero (bringing with it the fundamental natural frequency): The strut is at the point of incipient buckling. Beyond the critical value the strut takes up one of two symmetrically located equilibria. In terms of potential energy the system changes from a parabolic to a double-well form. To place this in a more concise mathematical context, we return to the total potential energy expression and include higher-order terms in the analysis. We assume a solution (buckling mode) of the form

$$w = Q \sin\frac{\pi x}{L}, \tag{4.17}$$

which can then be used to evaluate the strain and end-shortening energies to give

$$V = \frac{1}{2}EI\left(\frac{\pi}{L}\right)^4\frac{L}{2}Q^2 + \frac{1}{2}EI\left(\frac{\pi}{L}\right)^6\frac{L}{8}Q^4 + \cdots$$

$$- P\left(\frac{1}{2}\left(\frac{\pi}{L}\right)^2\frac{L}{2}Q^2 + \frac{1}{2}\left(\frac{\pi}{L}\right)^4\frac{3L}{32}Q^4 + \cdots\right), \tag{4.18}$$

where terms including the first nonlinear energy contribution have been retained. Now we see how the form of the potential energy relates to P to give the various shapes and stiffness functions introduced in the previous section (with Q equivalent to X). For example, if P is greater than the Euler critical value then the potential energy takes the form of two minima separated by a hilltop (i.e., exactly the twin-well form of potential energy underlying the cubic stiffness of Duffing's equation (4.1)). This description subsumes the standard linear case when $P = 0$ (for small deflections) and also the reduction in lateral linear stiffness when $P < P^c$. Chapter 11 of this book will look at the critically loaded case ($P = P^c$), which gives a purely cubic stiffness, and the appendix describes a continous panel with a thermally induced axial load subject to lateral acoustic excitation: Even in this case (where small geometric imperfections are included) the relation with a single degree-of-freedom-oscillator is evident (Murphy, Virgin, and Rizzi, 1997).

Nonlinear stiffness is encountered in structural mechanics in a variety of situations, including, for example, the snap-through of a shallow arch (Poston and Stewart, 1978). In general, the path followed in this book is not to use, say, axial load, P, to change nonlinearity (akin to *unfolding* in catastrophe theory); instead, the stiffness is set in a double-well configuration, and the magnitude of (external) forcing, $f(t)$, is used to explore nonlinear effects.

4.4 Lumped Parameters

Before moving on to consider a specific low-order mechanical system, we briefly outline another means of reducing a continuous system with distributed properties to an equivalent simpler system in which the system parameters have been *lumped* in some consistent way. This approach is popular in the modeling of linear dynamics systems in which there is considerable advantage in having a relatively simple dynamic model (e.g., in a design context). Section 4.3 described how a thin beam can be modeled by a partial differential equation and how variables could be separated to give uncoupled ordinary differential equations in space and time. If we are only interested in the temporal response at a specific location (for example, at the free end of a cantilever beam), it is possible to lump the mass and stiffness characteristics. There are a number of ways of doing this.

Consider the lateral vibration of a thin, elastic cantilever beam. The governing equation of motion, based on Euler–Bernoulli (thin beam) theory, is still given by Equation (4.9) with $c = P = 0$. Solving this equation after imposing the boundary conditions of zero deflection and slope at the clamped $(x = 0)$ end:

$$w = w' = 0 \tag{4.19}$$

and zero bending moment and shear at the free $(x = l)$ end:

$$w'' = w''' = 0, \tag{4.20}$$

results in the frequency equation

$$\cos(\beta_n l)\cosh(\beta_n l) + 1 = 0. \tag{4.21}$$

The lowest natural frequency is given by

$$\omega = (\beta l)^2 \sqrt{\frac{EI}{ml^4}} \tag{4.22}$$

with $\beta l = 1.875$, and a corresponding mode shape (the first bending mode with an arbitrary magnitude) is

$$w(x) = C\left[\sin(\beta x) - \sinh(\beta x) - \alpha\left(\cos(\beta x) - \cosh(\beta x)\right)\right], \tag{4.23}$$

where

$$\alpha = \left(\frac{\sin(\beta l) + \sinh(\beta l)}{\cos(\beta l) + \cosh(\beta l)}\right). \tag{4.24}$$

Similar expressions can also be obtained for the higher modes.

However, suppose we only need an approximate expression for natural frequency and the response at the free end. We can *lump* the distributed mass of the beam at the free end by equating kinetic energy, that is,

$$T = \frac{1}{2}M_e \dot{X}^2 \equiv \frac{1}{2}m \int_0^L \dot{w}^2\, dx, \tag{4.25}$$

where X is now the lateral deflection (and equivalent to the original response amplitude $w(l)$). Using an approximate deflected shape the equivalent lumped mass is found to be $M_e \approx 0.25ml$. A similar process, based on equating potential energy, can be used to extract an equivalent spring constant, $K_e = 3EI/l^3$ (a familiar result from engineering beam theory). Using these lumped characteristics we have

$$\omega_n = \sqrt{K_e/M_e}, \tag{4.26}$$

which upon evaluation results in a coefficent of 1.86 (compared with 1.875 from Equation 4.22). That only a relatively small proportion of the mass and stiffness should be concentrated at the free end makes intuitive sense. A variety of other distributed engineering components can be replaced by their lumped equivalents (including damping characteristics) and the results can be tabulated (Cochin and Plass, 1990).

The preceding discussion is related to the Rayleigh–Ritz method briefly touched upon in Section 4.3 (Thompson, 1982). Given the form of potential and kinetic energy we can assume mode shapes (satisfying the boundary conditions) and extract the natural frequencies from Rayleigh's quotient. The assumed mode shapes need not be extremely accurate (polynomials are a reasonable place to start) – an especially useful property given a relatively complicated geometry. Solutions tend to become more accurate as more terms in the assumed mode shapes are added. Galerkin's method attacks a similar class of problems using the method of weighted residuals (Meirovitch, 1997).

4.5 A Contrived Mechanical Paradigm

Despite the foregoing relevance of continuous beams, it is the compelling analogy between a point mass sliding on a curve under the action of gravity and a class of simple oscillator equations that is exploited throughout this book. The physical system is mostly based on attempting to mimic the dynamics of a system that lives in a double-well potential (Gottwald, Virgin, and Dowell, 1992; Moon, 1992; Shaw and Haddow, 1992; Gottlieb, 1997). In contrast to related standard problems in vibration, the overall dimensions are chosen to optimize, in some sense, the length- and time-scales for high data resolution. The concept of a roller

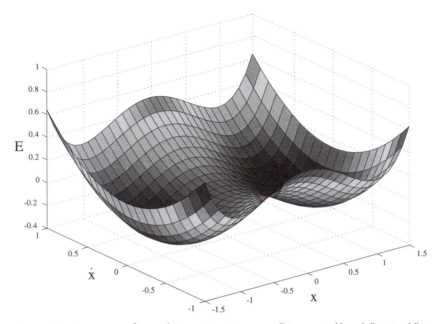

Figure 4.3: An energy surface with two minima (energy wells) separated by a hilltop (saddle).

coaster will thus be developed, and Figure 4.3 provides a further visual aid in this development as a surface with contours (horizontal cuts) representing levels of constant total energy, and the $\dot{x} = 0$ line described by Equation (4.5).

We will therefore focus entirely on a low-order system (with a single mechanical degree of freedom) using relatively unambiguous modeling to illustrate many interesting features in nonlinear oscillations. The ability to verify behavior experimentally is fundamental to this treatment. The experimental paradigm is designed with pedagogical intent to provide a relatively simple-to-measure, high-resolution, high signal-to-noise ratio system, requiring minimal filtering and signal conditioning.

In general, the system characteristics (e.g., mass, damping, and stiffness) will be fixed, and the forcing parameters will be used as the control parameters. The occurrence of instability at both local and global levels as these parameters are changed will thus provide the key evolution. This also gives a strong contrast with linear systems, which typically do not suffer bifurcations or sensitivity to initial conditions, and for which the distinction between local and global behavior has no meaning.

The progression of encountering new nonlinear features is thus generated by applying greater external energy to the system. Increased forcing will encourage the full spectrum of behavior, ranging from relatively mundane linear free vibration (at least from a topological point of view) to the fascinatingly complex world of nonlinear forced vibration with its display of chaos, fractal basin boundaries, and unpredictability.

Chapter 5

Mathematical Description

5.1 Introduction

At this point we wish to put some mathematical flesh on the intuitive bones of the preceding discussion. We shall consider how the general equations of motion and energy expressions are related for a point mass moving in a vertical plane, along a (single-valued) path described by $y = y(x)$ and under the influence of (vertical) gravity (see Figure 5.1).

Before embarking on a detailed Lagrangian description of the dynamics *in the horizontal projection*, we take a brief look at how the tangential motion along the curve is related to the governing equation of motion. We intuitively anticipate a local minimum to be associated with an equilibrium position, for a maximum to act, to some extent, as a confining barrier, and also for the speed of motion to be related to the local slope of the curve. Supposing we have a parabolic curve $y = x^2$, we might naively assume that the behavior along s can be described by a simple linear oscillator (the particle motion would clearly be periodic including a change in direction at the maximum excursion from equilibrium). But upon differentiation we have

$$dx = \frac{dy}{2\sqrt{y}}, \tag{5.1}$$

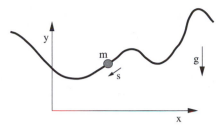

Figure 5.1: An arbitrary path $y = y(x)$ in a vertical gravitational field.

and using the relationship

$$ds^2 = dx^2 + dy^2, \tag{5.2}$$

we obtain

$$s = \int \sqrt{1 + 1/(4y)}\, dy, \tag{5.3}$$

which is clearly not the parabolic relation $s \approx \sqrt{y}$. It therefore seems reasonable to explore the extent to which the x coordinate of the position vector will mimic the behavior of a specific differential equation (Todd, 1996; Gottlieb, 1997).

For planar motion we can write

$$T = \frac{1}{2} m \mathbf{v}_{cm} \cdot \mathbf{v}_{cm}, \tag{5.4}$$

$$V = mgy, \tag{5.5}$$

where T and V are kinetic and potential energies, respectively, and $\mathbf{v}_{cm}$ is the velocity of the center of mass, m. This velocity vector can be written as

$$\mathbf{v}_{cm} = \dot{x}\hat{\mathbf{i}} + \dot{y}\hat{\mathbf{j}}, \tag{5.6}$$

with the magnitude given by $v = \sqrt{\dot{x}^2 + \dot{y}^2}$. But we also have the constraint $y = y(x)$, and since

$$\dot{y} = (dy/dx)\dot{x} \equiv y_x \dot{x}, \tag{5.7}$$

we can write the kinetic energy as

$$T = \frac{1}{2} m \left(1 + y_x^2\right) \dot{x}^2. \tag{5.8}$$

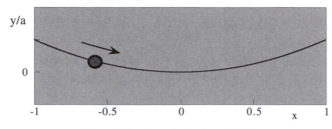

Figure 5.2: A parabolic shaped path.

A simple application of Lagrange's equation,

$$\frac{d}{dt}\left(\frac{\partial T}{\partial \dot{x}}\right) - \frac{\partial T}{\partial x} + \frac{\partial V}{\partial x} = 0, \tag{5.9}$$

yields the undamped (conservative) equation of motion describing the position as a function of time:

$$m\left(1 + y_x^2\right)\ddot{x} + m y_x y_{xx}\dot{x}^2 + mg y_x = 0. \tag{5.10}$$

If we now assume that the shape of the curve is quadratic, then $y = (1/2)ax^2$ as in Figure 5.2, and Equation (5.10) becomes the equation of motion governing the dynamics of the particle on the horizontal projection:

$$(1 + a^2x^2)\ddot{x} + a^2x\dot{x}^2 + gax = 0. \tag{5.11}$$

To what extent does this equation mimic the linear oscillator

$$m\ddot{x} + kx = 0, \tag{5.12}$$

which can be written in its familiar nondimensional form

$$\ddot{x} + \omega_n^2 x = 0? \tag{5.13}$$

We see that if a is relatively small such that a^2 can be neglected, then we recover the linear oscillator with the natural frequency given by $\omega_n = \sqrt{ga}$, and we see that the more shallow the curve, the more the point mass motion resembles a linear oscillator.

Clearly, certain assumptions have been made in this modeling process. Since this development is directed toward a macro-mechanical, roller-coaster type experimental configuration, it is natural to ask what are the influence of rigid-body effects, since a point mass is theoretically unrealizable. For a finite mass distribution we can now write the position vector as (Goldstein, 1980; Marion and Thornton, 1988)

$$\mathbf{r} = \mathbf{r}_{cm} + \mathbf{r}_{rel}, \tag{5.14}$$

where the two parts refer to the original (center of) mass with respect to the inertial frame and an additional term describes the position of any point on that body relative to the center of mass. This adds an additional term $(1/2)I_{cm}\dot\theta^2$ to the kinetic energy expression with a new motion variable $\theta = \theta(x)$. However, simple kinematics can be used to show that

$$\cot\theta = -\frac{dy}{dx}, \tag{5.15}$$

providing the finite length of the mass (cart) is negligible (Todd, 1996). A new application of Lagrange's equation now leads to

$$\left[m\left(1 + y_x^2\right) + I_{cm}\frac{y_{xx}^2}{\left(1 + y_x^2\right)^2} \right]\ddot{x}$$

$$+ \left[my_x y_{xx} + I_{cm}\left(\frac{y_x^2 y_{xx} y_{xxx} + y_{xx} y_{xxx} - 2y_x y_{xx}^3}{\left(1 + y_x^2\right)^3} \right) \right]\dot{x}^2$$

$$+ mgy_x = 0. \tag{5.16}$$

Before various damping mechanisms are considered, we also add the effect of moving the curve horizontally by a prescribed amount $u(t)$. Note that $y(t)$ was used for this purpose in Section 2.4 following convention, but here $u(t)$ is used so as not to be confused with the vertical coordinate. The generalized velocity vector expression now is given by

$$\mathbf{v}_{cm} = (\dot{x} + \dot{u})\hat{\mathbf{i}} + y_x\dot{x}\hat{\mathbf{j}}, \tag{5.17}$$

and again augmenting the kinetic energy and applying Lagrange's equation leads to the same expression as Equation (5.16) but now with $-m\ddot{u}$ on the right-hand side.

Again supposing we have the parabolic curve $y = (1/2)ax^2$, then the governing equation of motion becomes

$$\left[m(1 + a^2x^2) + I_{cm}\frac{a^2}{(1 + a^2x^2)^2} \right] \ddot{x}$$
$$+ \left[ma^2x + I_{cm}\left(\frac{-2xa^4}{(1 + a^3x^3)^3}\right) \right] \dot{x}^2 + mgax = -m\ddot{u}. \quad (5.18)$$

The analogy to the linear oscillator is still clear if the rotary inertia of the body about the center of mass (and hence I_{cm}) is relatively small. Furthermore, if we assume a harmonic motion of the curve itself, and of the form

$$u(t) = f\sin(\omega t + \phi), \quad (5.19)$$

then the right-hand side of Equation (5.18) becomes

$$f\omega^2 \sin(\omega t + \phi). \quad (5.20)$$

The extent to which these approximations are appropriate in modeling a typical mechanical experiment will be discussed later (Shaw and Haddow, 1992) where we include a discussion of an alternative approach based on arc length. Some interesting aspects of the inverse problem (i.e., given an equation of motion, what actual curve should be used to mimic it?) are discussed in Section 5.3.2 (Gottlieb, 1997).

5.2 Damping

We have already mentioned that in practice energy is inevitably dissipated and the particle would eventually come to rest at the bottom of the curve (in the absence of external forcing). Modeling the actual damping mechanism present in a physical system poses a notoriously difficult task and is influenced by a myriad of different factors including the area of contact, surface finish, lubrication, normal load, damage or wear, temperature, humidity, and "frictional memory" (den Hartog, 1930; Rabinowicz, 1959; Tabor, 1981; Steidel, 1989). The approaches most often used are somewhat empirical, and to capture the major dissipative effects relative to the dynamic behavior of interest, the two mechanisms of viscous and Coulomb damping will be considered. We shall focus on a periodically forced system with a single mechanical degree of freedom as described above. Even with these relatively simple damping models and confidence in their appropriateness, we will see later that, when they act simultaneously, it is difficult to measure their effects.

In mechanical engineering and applied physics, it is common practice to assess the role of (and measure) damping in terms of (i) the rate of decay of transient oscillations in free vibration, that is, an exponential decay for viscous damping (logarithmic decrement – see Section 2.5.1) and a linear decay for Coulomb damping, and (ii) the energy loss balanced by external excitation, for instance, the primary effect of damping in forced vibration is to limit the amplitude of response at resonance (leading to the half-power method for viscous damping – see Section 2.5.2), and the equivalent viscous damping (Thomson, 1981) for Coulomb friction (see Section 2.5.3). It is also worth pointing out here that sufficiently high Coulomb damping may also suppress motion altogether.

5.2.1 Viscous Damping

The most commonly assumed form of energy dissipation in mechanical systems is linear viscous damping, that is, a damping force that is directly proportional to velocity. This model has been shown to capture the majority of observed behavior in many practical circumstances, and its popularity has been enhanced by the fact that, in simple analytical studies, it does not add significantly to the algebraic complexity (e.g., superposition still holds).

However, in the present application we need to take account of the fact that it is a horizontal projection of the mass motion that is of interest here, and hence we need to consider the vectorial nature of this force. It is reasonable to assume that the damping force is a function of speed $\dot{s}$ in the tangential direction:

$$\vec{F}_{\text{damping}} = -\text{sgn}(\dot{s})c|\vec{v}|\hat{\mathbf{e}}_t = -\text{sgn}(\dot{s})c\dot{s}\hat{\mathbf{e}}_t, \tag{5.21}$$

where s is the arc-length coordinate defined in Equation (5.2), c is a damping coefficient and "sgn" is the signum function. Making the substitution $dy = y_x\,dx$ in Equation (5.2), dividing through by dt, and squaring both sides leads to

$$\dot{s}^2 = \left(1 + y_x^2\right) \dot{x}^2. \tag{5.22}$$

Rayleigh's dissipation function (see Section 2.2 and (Meirovitch, 1997)) can then be used:

$$P = \frac{1}{2}c\dot{s}^2 = \frac{1}{2}c\left[\left(1 + y_x^2\right)\dot{x}^2\right], \tag{5.23}$$

which, upon placement in

$$\vec{F} = \frac{-\partial P}{\partial \dot{x}}, \tag{5.24}$$

provides the damping force in the horizontal direction:

$$\vec{F}_{\text{viscous}} = -c\left(1 + y_x^2\right)\dot{x}\hat{\mathbf{i}}. \tag{5.25}$$

We see that this is not strictly a damping model in the usual sense since it is not solely a function of velocity, but it does provide a rational basis for positive energy dissipation (Todd, 1996).

An earlier analysis (Gottwald, Virgin, and Dowell, 1992) developed for the experimental system to be described later initially assumed a viscous damping model in which the damping force was proportional to $\dot{x}$ only. This was used in the simulations described in Chapters 7 and 8. This model was then replaced by Equation (5.25), which is used in simulations in Chapter 14. The difference between them is relatively minor (Todd, 1996).

5.2.2 Coulomb Damping

The presence of dry friction is also very common in a variety of mechanical engineering situations. Dry friction is inherently nonlinear (typically piecewise linear) and remains the subject of considerable discussion in the literature (Rabinowicz, 1959; Levitan, 1960; Tabor, 1981). Complex stick–slip behavior can occur and Coulomb damping may be sufficient to prevent any motion from occurring. Since the modeling of Coulomb damping is difficult, we will start by adopting a relatively simple model in which the static and kinetic coefficients are assumed equal (i.e., assuming the contact force is proportional to the static normal force) and then assess the role played by kinematics and geometry to obtain plausible energy dissipation terms in the governing equations of motion. Thus we envision the particle *sliding* along a wire and model the dominant dry friction effects on oscillatory behavior.

We assume a dissipation function for Coulomb damping in the form

$$P = \mu N \dot{s} = \mu N \sqrt{(1 + y_x^2)} |\dot{x}|, \tag{5.26}$$

where again $\dot{s}$ is the speed (assumed positive and tangential to the instantaneous velocity vector), N is the normal contact force, and μ is the kinetic coefficient of friction.

The static normal force N varies with $mg \cos \theta$ and therefore

$$N = \frac{mg}{\sqrt{1 + y_x^2}}, \tag{5.27}$$

and again using Equation (5.24) we obtain

$$\vec{F}_{\text{Coulomb}} = -\mu mg \frac{\dot{x}}{|\dot{x}|} \hat{\mathbf{i}}. \tag{5.28}$$

This is clearly a piecewise linear force that is independent of velocity (but not the direction of motion). This expression could also have been obtained from the principle of virtual work. We thus have a combined energy dissipation model of

$$\vec{F}_{\text{damping}} = -\left(c + \frac{\mu mg}{(1 + y_x^2) |\dot{x}|} \right) (1 + y_x^2) \dot{x} \hat{\mathbf{i}}. \tag{5.29}$$

Finally, incorporating this expression into the governing equation of motion results in

$$\left[1 + y_x^2 + (I_{\text{cm}}/m) \frac{y_{xx}^2}{(1 + y_x^2)^2} \right] \ddot{x} + \left(\frac{c}{m} + \frac{\mu g}{(1 + y_x^2) |\dot{x}|} \right) (1 + y_x^2) \dot{x}$$

$$+ \left[y_x y_{xx} + (I_{\text{cm}}/m) \left(\frac{y_x^2 y_{xx} y_{xxx} + y_{xx} y_{xxx} - 2 y_x y_{xx}^3}{(1 + y_x^3)^3} \right) \right] \dot{x}^2$$

$$+ g y_x = f \omega^2 \sin(\omega t + \phi), \tag{5.30}$$

where again a harmonic base displacement has been assumed.

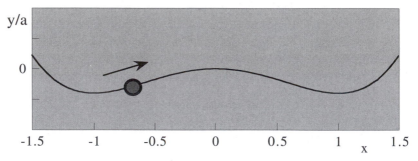

Figure 5.3: The double-well potential shape $(b/a=1)$.

Now, if we return to the assumption that the mass can be described as a single particle ($I_{cm} = 0$), then this equation reduces to

$$\left(1 + y_x^2\right)\left[\ddot{x} + \left(\frac{c}{m} + \frac{\mu g}{(1 + y_x^2)|\dot{x}|}\right)\dot{x}\right] + y_x y_{xx}\dot{x}^2 + g y_x$$

$$= f\omega^2 \sin(\omega t + \phi). \tag{5.31}$$

To obtain the equation of motion for a specific curve shape, the function $y = y(x)$ is placed into the above expression.

At this point we start to focus attention on the specific shape for the curve given by

$$y(x) = -\frac{1}{2}ax^2 + \frac{1}{4}bx^4, \tag{5.32}$$

with a view to incorporating this shape as a basis for mimicking Duffing's equation (see Figure 5.3). This form thus gives two symmetric local minima at $x = (a/b)^{1/2}$ and a local maximum at $x = 0$.

Placing this function for $y(x)$ into Equation (5.31) leads to (omitting damping)

$$\ddot{x}[1 + (-ax + bx^3)^2] + (-ax + bx^3)(-a + 3bx^2)\dot{x}^2$$

$$+ g(-ax + bx^3) = f\omega^2 \sin(\omega t + \phi). \tag{5.33}$$

We therefore have a second-order, ordinary differential equation with constant coefficients, but it remains to be seen how closely it mimics Duffing's equation.

5.3 Nondimensionalization

It is always advantageous to reduce a given physical problem to a non-dimensional form; for example, the stiffness and mass are most effectively viewed as contributions to the natural frequency and thus directly related to this important universal attribute of the response. It is natural to nondimensionalize a problem in time and space as a starting point. The obvious time scaling is to use the natural frequency of small (linear) oscillations:

$$\tau = \omega_n t. \tag{5.34}$$

With regard to the double-well shape just introduced, the natural frequency can be obtained by linearizing about one of the stable (static) equilibrium positions $x_e = \pm(a/b)^{1/2}$ to give

$$\omega_n^2 = 2ga. \tag{5.35}$$

In terms of the position scaling it is a natural choice here (although somewhat arbitrary) to scale the position x by the horizontally projected distance between the origin and the stable equilibria, that is,

$$X = x/x_e. \tag{5.36}$$

These then provide a means of scaling the forcing parameters: $\Omega = \omega/\omega_n$ and $F = f/x_e$. An important new geometric nondimensional parameter is apparent:

$$\alpha = \frac{\omega_n^2 x_e}{2g}. \tag{5.37}$$

This parameter has a strong influence on the form of the governing equation of motion.

Finally, incorporating the harmonic forcing expression and damping mechanisms (with ζ defined in Section 2.3.1) leads to the complete nondimensional equation of motion (from Equation 5.33):

$$[1 + \alpha^2 X^2 (X^2 - 1)^2](X'' + 2\zeta X') + \frac{\mu}{2\alpha} \operatorname{sgn}(X')$$

$$+ \alpha^2 X'^2 X (X^2 - 1)(3X^2 - 1) + \frac{1}{2} X(X^2 - 1)$$

$$= F\Omega^2 \sin(\Omega\tau + \phi), \tag{5.38}$$

where a prime denotes differentiation with respect to nondimensional time, τ. In the results described in subsequent chapters the parameters will generally be given in their nondimensional forms. The exception to this is time, which is generally left in dimensional units (seconds) to provide the reader with a reference point, and some of the power spectra are left in hertz for the same reason.

We are now able to make a general comparison with Duffing's equation (4.1). The first thing to notice is that the oscillator belongs to the class of transmissibly forced systems, behaving in a similar way to a building frame responding to the imposition of a foundation motion, for example, during an earthquake (see Figure 2.6(b)). That is, the effective amplitude of the external force is also dependent on its frequency and we see the similarity between this system and the standard (direct mass excitation case) when the system is forced at resonance ($\Omega = 1$). The presence of the Coulomb friction term adds a complication, although in most practical cases it can be considered to be relatively small, especially in the resonant, relatively large-amplitude regimes of particular interest throughout this book. The presence of the factor 0.5 in front of the linear stiffness term is a consequence of the 0.5 factor in the definition of α and was included for the following reason. Conducting a linearization about either of the two stable equilibria (see Section 3.5.1) will result in a stiffness of unity; that is, we effectively scale the local potential energy (and hence local natural frequency) to be the same as the linear oscillator. Finally, if the parameter α can be taken as being small, then this equation begins to resemble Duffing's equation quite strongly. An experimental mimicry of Duffing's equation will thus be more successful if a shallow curve is used with a small (concentrated) mass and minimal friction. In the subsequent chapters, where a direct comparison is made between experimental data and solutions to the governing equations, the full equation of motion (5.38) will be solved (numerically) unless otherwise stated. We also mention at this point that a simpler damping model is used in some of the studies to be described (i.e., following from the discussion in Section 3.2.1, the viscous damping term is assumed to be independent of the first bracketed term in Equation 5.38). Again, given the relatively light damping in typical mechanical systems, this is a minor change.

5.4 Alternative Analytic Approaches

Before leaving the question of mathematical modeling we briefly consider a couple of other perspectives. Since the focus in this book is on qualitative behavior, the analogy between a track–cart mechanical system and certain nonlinear oscillators (and Duffing's equation in particular) is strong provided the parameter α in Equation (5.38) is small, and even if this is not the case, there is still a broad qualitative similarity in the dynamic response. However, perhaps it is possible to choose a specific form of track shape, y, such that these extra (nonlinear) terms are effectively minimized or even eliminated. Thus it is intriguing to consider the inverse problem: Given a desired form of equation, what shape should the track take to exactly mimic the desired equation? The following sections give some alternative approaches to this problem.

5.4.1 The Arc-Length Coordinate

Returning to the basic modeling of the cart–track system it is instructive to consider a more direct analytic formulation based on the arc-length coordinate. As pointed out by Shaw and Haddow (Shaw and Haddow, 1992) this is very much related to the classic tautochrone problem. For simplicity, damping and external forcing will be ignored and the specific case of a linear oscillator and its relation to a parabolic track will be discussed. In a practical sense, the arc-length coordinate would not be as easy to use in the actual construction of a track, although, of course, it is the arc-length motion that is measured directly prior to any calibration to extract the horizontal projection.

Consider again the system shown schematically in Figure 5.2 where we assume $a = 1$ for simplicity. We can write the kinetic energy in terms of the arc length s as

$$T = \frac{1}{2}\delta\dot{s}^2, \tag{5.39}$$

where $\delta = m + I/r^2$ (the effective translational inertia). The potential energy is given by

$$V = mgH(s), \tag{5.40}$$

where $H(s)$ is the vertical height of the track as a function of the arc-length. Lagrange's equations can be used to obtain the linear oscillator

$$\ddot{s} + (mg/\delta)H_s = 0, \qquad (5.41)$$

where $H_s = \partial H(s)/\partial s$. Hence, the desired form of oscillator can be prescribed.

Clearly, a change of coordinates (from s to x) provides the appropriate track shape in terms that would facilitate the track construction and provide a comparison with earlier parts of this chapter. It can be shown that the relevant relations are given by (Shaw and Haddow, 1992)

$$x - x_0 = \int_{s_0}^{s} \left[1 - H_s^2\right]^{1/2} ds \qquad (5.42)$$

and

$$s - s_0 = \int_{x_0}^{x} \left[1 + h_x^2\right]^{1/2} dx, \qquad (5.43)$$

where $h_x = [\partial h(x)/\partial x]$. These provide the relation between $h(x)$ and $H(s)$. We also observe the physical constraint $|H_s| \le 1$ corresponding to a vertical tangency.

For the linear oscillator (Equation 5.41) with $mg/\delta = 1$ we have

$$H(s) = s^2/2, \qquad (5.44)$$

and evaluation of

$$\int_{s_0}^{s} [1 - s^2]^{1/2} ds, \qquad (5.45)$$

with $s_0 = x_0 = 0$, results in

$$x = \frac{1}{2}\left[s\sqrt{1 - s^2} + \sin^{-1} s\right]. \qquad (5.46)$$

This transformation is valid over the interval $-1 \le s \le +1$, and hence $-\pi/4 \le x \le +\pi/4$. This expression is close to being a straight line, especially for small s.

Now, using the inverse transformation based on Equation (5.43) (achieved in practice by reflecting Equation 5.46 about the $s = x$ line), the expression for $h(x)$ can be evaluated (although not in an explicit

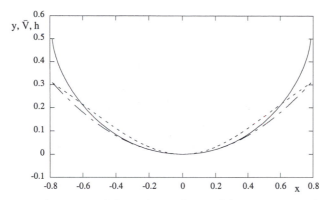

Figure 5.4: Some alternative track shapes that result in simple harmonic motion: The continuous curve, $h(x)$, is based on the arc-length coordinate, s. The dashed curve, $\bar{V}(x)$, is based on a potential that produces a motion isodynamical to a linear response (for a given amplitude of motion). The dot–dashed curve, $y(x)$, is the parabolic track shape.

form). This is shown in Figure 5.4, together with the purely parabolic track. It can be seen that the curves remain close for small-amplitude motion but begin to diverge as the slope becomes steeper – a result anticipated by the desirability of a shallow track as discussed in the opening section of this chapter.

5.4.2 Isodynamical Systems

Another approach to establishing the relation between a one-dimensional potential and the motion of a bead on a wire in a vertical gravitational field is discussed by Gottlieb (Gottlieb, 1997). He found that an isodynamical correspondence between the track shape and potential energy resulted in the algebraic condition

$$\bar{V}(x) = \frac{y + y(A)y_x^2}{1 + y_x^2}, \tag{5.47}$$

where $\bar{V}(x) = V(x)/(mg)$. An explicit dependence on the amplitude of motion, A, thus appears. For the parabolic track equation

$$y(x) = \frac{1}{2}x^2 \tag{5.48}$$

and assuming $A = \pi/4$, the potential isodynamically equivalent to a parabolic track shape is given by

$$\bar{V}(x) = \frac{0.808x^2}{1 + x^2}. \tag{5.49}$$

This is also shown in Figure 5.4. In comparing these shapes it is important to realize that a particle moving on the isodynamical curve behaves (on the horizontal projection) in an equivalent time-dependent way over the specific range of motion shown by the extent of the horizontal axis (i.e., on the interval $[-\pi/4, \pi/4]$).

It can also be shown that various mathematical subtleties may occur under some circumstances (e.g., nonunique solutions, and even kinks in the curves). Additional results on the double-well Duffing potential again confirm the desirability of relatively low slopes. The modeling of damping (including Coulomb friction), external forcing, and any rotational inertia effects (violating the point mass assumption) would provide considerable complications in applying these methods. Again, we emphasize that the specific form of experimental system considered in detail for much of the remainder of this book provides a compelling qualitative analogy with the prototypical nonlinear oscillator, namely Duffing's equation.

It is always possible to refine modeling (e.g., we chose to ignore the rotary inertia effects (Section 3.1)). There may be some (small) draglike damping on the finite-size mass caused by air resistance (hence leading to a velocity-squared term in the governing equations), or a small amount of out-of-plane motion, or a small but unavoidable asymmetry, or even temperature effects leading to nonstationarity. However, the behavior to be presented will show a good correlation between theoretical and experimental results, that is, the fundamental qualitative nonlinear behavior is captured.

Chapter 6

The Experimental Model

6.1 Introduction

We have seen that in order to design a close mechanical analogue of the equation of motion (developed in the previous chapter) a number of objectives have to be achieved. An initial approach might consist of allowing a small ball to roll on a grooved guide (Marion and Thornton, 1988). However, it is then difficult to monitor the motion of the ball (at least in an accurate manner) and significant slippage can occur. Suppose we wish to build a small cart or roller coaster. In this case it is relatively easy to measure the position of the cart based on the output of a rotational potentiometer attached to an axle. The rotary inertia of the cart can be minimized by keeping the cart small, and the α term can also be adjusted through the track geometry. To avoid slippage between the cart wheels and the track, a chain–sprocket system can be used. This does have the drawback of complicating the damping modeling but it will be seen that the overall damping in the model is quite small. Another advantage of the chain–sprocket guide is that it minimizes the possibility of the cart actually leaving the track during fast motions (Gottwald, Virgin, and Dowell, 1992).

We will also see that to replicate Duffing's equation it is desirable to have a relatively shallow curve so as to minimize those nonlinear terms in the accurate equation of motion (5.38) that do not appear in the standard

Duffing's equation (4.1). We do not wish to stray too far from the familiar form of Duffing's equation, although we realize that typical nonlinear features are by no means restricted to a narrow class of ordinary differential equations. The features to be described are actually quite generic and robust. The theoretical development in the previous chapter showed that these effects can be grouped together by consideration of a single nondimensional parameter α in the experiment. Equation (5.37) showed that α can be made smaller by reducing the vertical distance between the unstable equilibrium (hilltop) and the symmetrically positioned stable equilibria to reduce the vertical component of the acceleration. We also see that increasing the natural period will tend to reduce α and this is achieved by making the track more shallow; that is, we make the parameter a small to effectively bring down the height of the potential energy at the origin. Since the linear natural frequency is based on oscillations within one of the wells, we see that the parameter b does not directly affect this scaling.

The other deviations from Duffing's equation can be minimized by ensuring a relatively small mass (to reduce I_{cm}), minimal damping, and forcing in the vicinity of main resonance ($\Omega \approx 1$) to more closely approximate direct mass excitation. We again emphasize that since most of the computations are conducted on the "full" equation of motion (5.38), these scaling efforts are made simply to keep Duffing's equation as a meaningful analogy.

6.2 The Model

With the desirable features from the previous chapter in mind, we develop the following experimental mechanical model (as shown in Figure 6.1).

Two parallel Plexiglas sheets were machined according to Equation (5.32), and accurately bolted together 0.047 m apart. Both sheets were finely shaped from a plywood template and surface irregularities were minimized. The horizontal distance between the hilltop at the origin and the position of the two symmetric minima is 0.273 m (this will provide the basic *unit* for position nondimensionalization), and the vertical distance is 0.07 m. Hence the actual shape of the curve (in meters) is given by

$$y(x) = -\frac{1}{2}(3.76)x^2 + \frac{1}{4}(50.4)x^4. \tag{6.1}$$

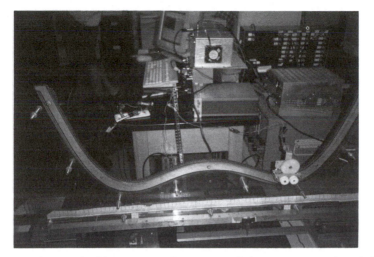

Figure 6.1: Photograph of the experimental apparatus with the "cart" at rest in the right-hand well.

According to the preceding theory, the linearized natural frequency of the system is $\omega_n^2 = 2ga$, and we expect a measurement close to $\omega_n = 8.6$ rad/s ($= 1.37$ Hz). It turns out that the measured natural frequency tended to be somewhat less than this value, a discrepancy most likely due to the influence of the chain–sprocket antislippage mechanism to be described later and also to some of the other modeling issues discussed in Chapter 5. Furthermore, we can estimate the nondimensional parameter α directly from the track geometry to give $\alpha = 1.03$, which is not small, but again the lower measured natural frequency results in a lower nondimensional parameter $\alpha \approx 0.77$, and we see from Equation (5.38) that this parameter generally enters as a squared contribution.

The *point mass* consists of a short-wheelbase cart. This includes Teflon wheels with small grooves (to reduce damping), which run along the track and incorporate low-friction roller bearings. A lead ballast provides a lower center of gravity of the vehicle (relative to the track surface) such that the rotary inertia is minimized. This also has the advantage of lowering the effective damping. The weight of the cart is approximately 1 kg. The antislip mechanism consists of a chain–sprocket combination running along the length of the track to confine any slippage to one chain link. This effect was measured to be no greater than 0.005 units (the definition of one *unit* is the horizontal distance between the maximum and either of the two adjacent minima). The antislippage system was introduced because initial trials showed that a small amount of sliding occurred. Although there was only a small amount per period of oscillation it was found that cumulatively the error would build up quite significantly, an especially undesirable situation when trying to characterize chaos, for instance. Again, the emphasis throughout this book is on qualitative behavior rather than on an exact theoretical–experimental correlation per se (Carlson and Gisser, 1981; Hilborn, 1994)

6.2.1 The Starting Gate

The free oscillations to be described in Chapter 7 were generated by fixing the initial position of the cart at a prescribed location using a *starting gate* and zero initial velocity. This device was attached to the track using a simple pinning arrangement and was operated by an electrically driven solenoid trigger. Thus a variety of initial positions starting from rest could be investigated. A more sophisticated method of investigating initial conditions in forced vibration problems allows the generation of random starting positions *and* nonzero velocities. Further discussion of this approach is deferred to Chapter 14.

6.2.2 The Impact Barrier

Chapter 12 will focus attention on the cart on a parabolic track with the addition of contact with a rigid *impact* barrier. This moveable barrier, consisting of a solid steel support frame and incorporating a stiff helical spring (located at the same level as the center of mass of the cart), could be bolted to the track at various locations. The stiffness of the spring was sufficiently high that the cart experienced a hard rebound, which can be modeled to reasonable accuracy by a coefficient of restitution. A more advanced analysis allows for a finite stiffness of the impact penetration and hence the consideration of a bilinear stiffness.

6.3 The Forcing Mechanism

To periodically excite the system, a forcing mechanism was designed. A naive approach to imparting a harmonic base motion might be to use a

Figure 6.2: Photograph of the Scotch-yoke forcing mechanism. The frequency is controlled by a variable-speed motor and the amplitude by a stepper motor.

slider-crank mechanism. Although this does produce a periodic input, it is not sinusoidal and, in fact, since it is the acceleration that is important, the deviation from a pure harmonic is magnified.

An appropriate mechanism is the Scotch-yoke driven shake table, as shown in Figure 6.2. The apparatus is capable of moving the track up to 0.15 m peak-to-peak (a little more than half a track unit) at frequencies up to about 2 Hz. Since one of the major goals in this experiment was to observe resonant behavior, this maximum frequency is somewhat greater than the natural frequency of the system. The amplitude is controlled using a stepper motor with the frequency controlled using a standard inductive motor. Figure 6.3 shows a representative trace (at 1 Hz) of the measured base motion as a time series, which was obtained from a potentiometer mounted to the underside of the shake table. Also shown is a power spectrum based on an FFT. The vertical scale in the power spectrum is decibels and hence the higher harmonics have considerably less energy. Plotting the base output against itself one-half cycle later showed a line with slope $= -1$ and zero intercept (Hilborn, 1994). The other components in the FFT are the inevitable result of experimental tolerances in the mechanism design (e.g., the gearing system, belt slippage, and looseness).

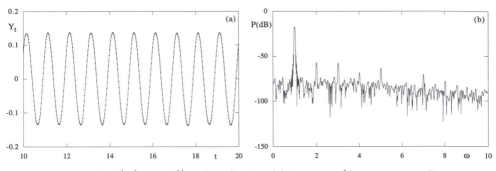

Figure 6.3: The horizontal base (input) motion. (a) Time series; (b) power spectrum (Frequency in Hz).

Because the Scotch-yoke is not well-suited to two-frequency excitation a hydraulically activated shake table is used to generate the quasi-periodic behavior to be described in Chapter 13. This facility consists of a 1.2 m by 1.2 m table, driven by a 50 kN hydraulic actuator, and a 10 gpm, 90 Hz servovalve. A variety of base excitations can be prescribed, including earthquake records, as input.

6.4 Data Acquisition

The past few years have seen a remarkable growth in computer-based data acquisition and analysis systems. One of the leaders in this area is the virtual instrument-based system of LabVIEW (National Instruments). This object-oriented programming environment simulates instruments but exists purely as software. Combined with the user-friendly Mac or PC systems it provides a powerful experimental platform.

For the measurement of the cart position, a multiturn potentiometer attached to one of the axles provided a continuous voltage proportional to the arc length of the track. This position can then be converted to a horizontally projected position (X) using a polynomial least squares fit:

$$X = A_1 V + A_3 V^3. \tag{6.2}$$

The values $A_1 = 0.32$, $A_3 = -0.0018$ were extracted from a curve fit (an odd polynomial was used owing to assumed symmetry of the track).

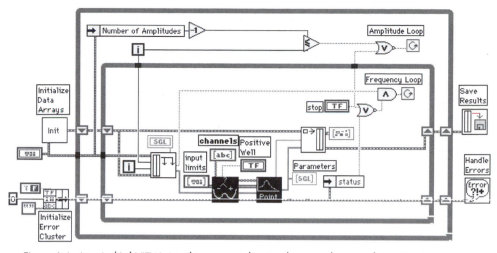

Figure 6.4: A typical LabVIEW virtual instrument showing the general approach to extracting a bifurcation diagram.

These coefficients are representative; occasionally these values were slightly adjusted to account for small alterations in the experiment.

Over the range $-1.6 < X < 1.6$, the measured voltage fell within the values $-7V$ to $+7V$, allowing the computer data acquisition system to measure over 2,800 cart positions with an average accuracy of about 0.001 track units (about 0.25 mm). A typical response was characterized by either 50 or 100 samples per period of oscillation.

Figure 6.4 shows a typical virtual instrument. This was set up to acquire and plot a response or bifurcation diagram (e.g., the response measured as a function of frequency). The versatility of this kind of modular system (within software) provides a very powerful and flexible basis for conducting the types of experiment described throughout this book.

6.5 Time-Lag Embedding

Because this system has a three-dimensional phase space, the position provides just one of the state variables. However, the forcing phase can be considered as another state variable (the cyclic symmetry of the forcing provides a further dimension). Given the fact that we are primarily interested in measuring the state variables of position and velocity (and their Poincaré sampling) a number of measurement choices present themselves. Clearly, a direct measurement is one option. It is possible to measure the velocity of the cart independently of position. The velocity can also be extracted from an acceleration time series by numerical (or electronic) integration. But the following approaches were used based on manipulating the position time series, which was relatively easy to measure using the method detailed earlier. Since the velocity is the time derivative of the position, a finite difference approach based on discrete data can be established by noting that the approximate (average) expression for velocity is

$$\dot{X} \approx \frac{1}{2\Delta t}(X_{i+1} - X_{i-1}), \tag{6.3}$$

where Δt is the time step. There is a vast literature on various more sophisticated ways of extracting rates of change of discrete data (Press et al., 1992). However, this process of numerical differentiation tends to accentuate the low-amplitude, high-frequency noise present in a signal (this is related to one of the reasons why a slider-crank was rejected for the forcing mechanism). A few phase projections will be shown later in which the velocity is based on this technique.

However, in much of the subsequent consideration of this system, use is made of time-lag embedding based on a reconstruction of an attractor in phase space. Generally the position is plotted against itself one quarter of a cycle later as a topologically equivalent phase space reconstruction (Liebert and Schuster, 1989; Casdagli et al., 1991; Ott, 1993). Clearly some velocity information is contained in two position measurements taken over a time lag, say δ (see Section 6.3). This concept can be generalized such that an N-dimensional state space orbit from a single observable variable $x(t)$ can be reconstructed using a set of single state variable measurements $\mathbf{x}(t) = [x(t), x(t+\delta), x(t+2\delta), \ldots, x(t+(N-1)\delta)]^T$. The choice of one-quarter cycle for δ represents a compromise that allows enough of the system dynamic evolution but not too much. For a harmonic oscillation, a quarter-cycle lag reproduces the velocity exactly. For highly complex systems, a delay based on the first zero crossing of the autocorrelation function has been suggested (Casdagli et al., 1991).

Clearly, the embedding dimension, N, should be greater than the actual (unknown) dimension of the system and certainly large enough so that reconstructed trajectories do not cross (from the No-Intersection theorem). A proposed rule (Ott, 1993) suggests using a dimension of at least twice the size of the original observed motion (plus one). But since this is generally unknown some other techniques have been developed, including using Karhunen–Loeve (K–L) decomposition (Broomhead and King, 1986), false nearest neighbors (Kennel, Brown, and Abarbanel, 1992), and singular value decomposition (Banbrook, Ushaw, and McLaughlin, 1997), which suggest that embedding (which need not be as high as the estimate above) and associated signal processing can be useful in noise reduction. In this book the focus is on low-order systems with low noise, and hence the dimension (generally three or less) is quite clear.

6.6 Measurement of Damping

Earlier chapters gave some theoretical consideration to the modeling of energy dissipation, including a number of ways of relating damping estimates to the metrics of typical experimental output. Even if the mechanisms of damping were known, it still remains a challenging task to measure these effects in an experiment, not least because various mechanisms may be acting simultaneously. Some recent theoretical work has suggested an approach for the simultaneous extraction of both viscous and Coulomb damping effects (Feeney and Liang, 1996). The approach adopted here consists of trying a variety of tests in order to determine a range of damping coefficients that capture the essential features of the behavior. Modeling and measuring damping in a mechanical system often presents a challenging task. Again, we emphasize that similar qualitative features occur over a range of parameter values, and hence a strict quantitative comparison between theory and experiment is not crucial to the material developed in this book.

Given the two energy dissipation mechanisms of viscous and Coulomb damping, the techniques outlined earlier were used in a variety of situations to obtain appropriate damping coefficients. Figure 6.5 shows a typical free decay generated by releasing the cart from an initial position away from the static equilibrium position in the left-hand well. It can be seen that initially there is a marked asymmetry in the response. We also expect an exponential decay for a viscously damped system and a linear decay for a Coulomb damped system. From the small-amplitude decay, a viscous damping coefficient in the range $0.01 < \zeta < 0.02$ was

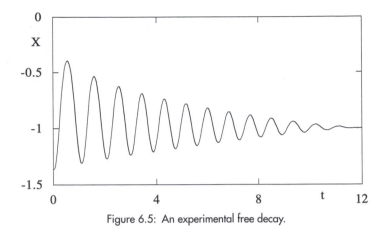

Figure 6.5: An experimental free decay.

estimated using logarithmic decrement (see Section 2.5.1). This range was confirmed by the sharpness of the resonant peak in forced tests (the quality factor – see Section 2.5.2). However, this approach assumes purely viscous damping. To estimate the Coulomb damping present in the system, the cart was placed on a flat track identical to the Duffing track. As the angle of tilt was increased, the slope of the track (β) that initiated motion (with a constant velocity) was noted and the damping coefficient (μ) extracted according to $\mu = \tan \beta$, to give an appropriate range $0.01 < \mu < 0.02$ (Bayly et al., 1994). Although this range appears very small (clearly the cart does not actually slide on the track) it compares favorably with results found using a forced test, an energy balance, and a number of parameter studies. In fact, this level of Coulomb damping may be negligible in large-amplitude resonant motion, but it is easy enough to retain it in the numerical simulations.

Certain refinements were made to the experimental system over an extended period of time that had the effect of altering certain measured properties (e.g., from time to time it was necessary to make adjustments to the antislippage and gear mechanisms thus requiring recalibration and, hence, a small adjustment to the measured damping values). For example, over the period during which experimental data were collected (a number of years), in addition to subtle changes in the damping, the natural frequency was measued within the range $1.15 < \omega_n < 1.22$. This then had the effect of altering the nondimensional parameter α over the range $0.72 < \alpha < 0.81$. However, any deviation from the baseline characteristics will be mentioned in the text as they occur. Also, the majority of the results to be presented will be in terms of nondimensional parameters, although time will typically be left in dimensional units (seconds) in most subsequent plots.

6.7 Poincaré Sampling

Poincaré mapping, first introduced in Section 3.3, is a stroboscopic sampling technique that relates flows and maps and is a fundamental tool in the investigation of nonlinear systems. An N-dimensional continuous system is thus reduced to an $(N - 1)$-dimensional discrete mapping. This is especially useful in the context of forced nonlinear oscillators since the significant trajectory evolution can now be studied in the plane, and because the external excitation is periodic, the forcing phase is an appropriate trigger for defining the *surface of section*. Hence, the trajectory states are sampled by their intersection with a plane (the Poincaré section) defined by a constant forcing phase.

In order to accomplish this experimentally, a number of alternative techniques have been employed. The approach finally adopted follows: To generate a pulse once every forcing cycle, small DC magnets (two per forcing cycle in the case of time-lag embedding) were placed directly onto the flywheel of the forcing mechanism. As the magnets passed a pickup, a pulse was generated within LabVIEW and position data points were extracted. This approach was also used to generate bifurcation diagrams for which the driving motor would step through the control parameter (usually frequency) at a relatively slow rate.

In the general forced vibration case we can then conveniently study the main features of the response by converting the three-dimensional flow (position $x(t)$, velocity $\dot{x}(t)$, phase $\phi \equiv \omega t$) into a two-dimensional map (sampled position $x_p(t)$ and velocity $\dot{x}_p(t)$, where p stands for Poincaré, and ϕ determines the sampling, i.e., once per forcing cycle) and also using delay coordinates to obtain a topologically equivalent two-dimensional mapping $(x_p(t), x_p(t + \delta))$, where δ is based on a quarter-cycle delay. Clearly, measuring one state at two discrete times is generally easier to accomplish than measuring three states over continuous time.

6.8 Other Experimental Issues

6.8.1 Signal Conditioning

Often, in the detection and acquisition of a mechanical quantity, it is necessary to modify a signal in some way, generally through the use of electronic manipulation. These include various amplifications to increase signal voltages or filtering to remove unwanted frequencies. Instrumentation circuitry is a vast field (Doeblin, 1990).

The data acquisition described earlier was used to measure the relative position of the cart to track location. The large scale of the experiment and low-order nature of the response meant that signal conditioning could be kept to a minimum. For power spectra computation, standard techniques were used to mitigate against aliasing, leakage, and noise, and the use of time-lag embedding obviated the need to measure all the state variables, as noted previously.

In most of the experimental runs the sampling rate was set at approximately 20 ms (50 Hz) or 6 ms (167 Hz), depending on the complexity of the motion. Many of the results presented later were not subject to filtering at all, although sometimes a low-pass Butterworth second-order filter (10 Hz) was used in certain circumstancs. Prior to power spectra analysis, time series were typically subject to a Hanning window (see Section 3.3), in which the data time set was subject to a convolution with the windowing function:

$$w_j = \frac{1}{2\Delta t} \left[1 - \cos \frac{2\pi j}{N - 1} \right], \tag{6.4}$$

that is, a function that is high in the middle and low on the sides. The choice of this specific form is not crucial, since the goal of reducing leakage can be achieved using various functions. Generally the length of data for FFT analysis was either set at 2^n directly or padded up to this value using zeros. More details of these signal conditioning techniques can be found in Refs. (Oppenheim and Schafer, 1975; Bendat and Piersol, 1986). For more complex signals (and higher-order systems) than the ones encountered in this book, the experimentalist must be careful that filtering does not significantly change the dynamics.

Furthermore, in response to the question: how good are the data? (Beckwith, Marangoni, and Lienhard, 1993), the experimentalist typically assesses the meaning of the word *good* in terms of the evaluation of precision and bias error and the statistics of samples and uncertainty. Although these issues are undoubtedly important, in this book, we take a rather straightforward meaning to the assessment of good data: The multitude of responses displaying close correlation between theory and experiment is considered confirmation that the data are of sufficiently high quality. Despite the relatively simple nature of the experimental systems and the sophistication of the modeling, the emphasis throughout is on the exhibition of certain *qualitative* behavior.

6.8.2 System Identification

The approach adopted in this book is to develop mathematical models of the experimental systems from a direct application of (generally) Lagrange's equation. The parameters of the governing equations of motion are then either measured directly (e.g., weighing a mass), indirectly (e.g., determining the stiffness from the natural frequency with a priori knowledge of the mass), or using some other commonly used technique (e.g., the log dec method; see Chapter 2) to estimate damping.

An alternative approach is based on system identification or parameter estimation (Yun and Shinozuka, 1980; Yasuda and Kamiya, 1990; Kesaraju and Noah, 1994). An assortment of techniques have been developed where a mathematical form of the system (typically the differential equation of motion) is assumed and then the coefficients are identified (fitted) from data. This approach has proved to be particularly successful for linear systems and is routinely incorporated as a component in the practical application of control (Goodwin and Payne, 1977; Ebert, 1984; Stengel, 1986). It has also been embraced by the nonlinear dynamics and chaos research communities since extracting qualitative dynamics from measured time series has universal application (Abarbanel et al., 1993). However, because this book is concerned with relatively well-defined, but nonlinear, single-degree-of-freedom systems, this approach is not pursued here.

Chapter 7

Free Oscillations

7.1 Introduction

We initiate this chapter by presenting a brief summary of the unforced response of the Duffing system, where the cart is given some initial energy (typically a nonequilibrium position), and the ensuing motion decays naturally. Some nonlinear free vibration responses will highlight the dependence on initial conditions. This will serve as a gentle introduction to the issue of predictability prior to full immersion in the extreme sensitivity to initial conditions to be considered later.

We shall start by looking at the free response of

$$\ddot{x} + \beta\dot{x} + ax + bx^3 = 0, \tag{7.1}$$

paying particular attention to the role of initial conditions. Suppose $a = -1, b = +1$, and thus the three equilibrium positions are given by $x_e = 0, \pm 1$. Furthermore, if we neglect damping for the moment ($\beta = 0$), we can consider contours of total constant energy based on the Hamiltonian:

$$H = \frac{1}{2}\left[\dot{x}^2 - x^2 + \frac{1}{2}x^4\right], \tag{7.2}$$

which is plotted for a range of H values in Figure 7.1. In practice these undamped phase trajectories are obtained by noting that $\ddot{x} \equiv \dot{x}(d\dot{x}/dx)$,

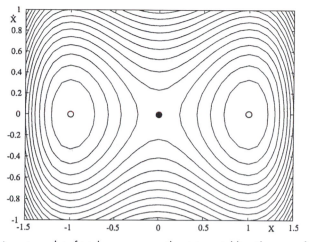

Figure 7.1: A contour plot of total energy versus the state variables. The state of the systems follows these contours in a clockwise direction.

separating variables, and integrating. We see nearly elliptical solutions (centers) if x remains in the vicinity of ± 1 and a saddle at the origin. This saddle is the common point of two homoclinic orbits, which separate bounded (single-well) and unbounded (cross-well) motion. The system is globally bounded. The equation of the separatrix corresponds to $H = 0$ and is given by

$$\dot{x} = \pm\sqrt{x^2 - \frac{1}{2}x^4},\qquad(7.3)$$

with a (forwards and backwards) time evolution of the form

$$x(t) = \pm\sqrt{2}\,\mathrm{sech}\,t.\qquad(7.4)$$

Since we will be initially interested in how the motion changes from essentially linear behavior, in the neighborhood of an equilibrium state x_e, it is appropriate at this point to conduct a simple linearization of Equation (7.1) by substituting $x = x_e + \delta$ to give (see Section 3.5.1)

$$\ddot{\delta} + \beta\dot{\delta} + \delta(3x_e^2 - 1) = 0,\qquad(7.5)$$

assuming that δ is small. Substituting in the three values for x_e gives

$$\ddot{\delta} + \beta\dot{\delta} + 2\delta = 0\qquad(7.6)$$

for motion about $x_e = \pm 1$ and

$$\ddot{\delta} + \beta\dot{\delta} - \delta = 0 \tag{7.7}$$

for motion about $x_e = 0$. The extent to which δ can be considered small is therefore dependent on the motion, which depends on the initial conditions.

It is quite clear from the form of Equations (7.6) and (7.7) how the state of the system evolves in time. It is convenient to relate this back to the characteristic eigenvalues (CEs) introduced earlier. For the system given by Equation (7.6) we can write the state matrix as

$$\begin{bmatrix} 0 & 1 \\ -2 & -\beta \end{bmatrix} \tag{7.8}$$

and with small positive damping we have eigenvalues with negative real parts, $\lambda_{1,2} = (1/2)(-\beta \pm \sqrt{\beta^2 - 8})$, and thus a stable classification. With Equation (7.7) we have

$$\begin{bmatrix} 0 & 1 \\ 1 & -\beta \end{bmatrix} \tag{7.9}$$

whose eigenvalues, $\lambda_{1,2} = (1/2)(-\beta \pm \sqrt{\beta^2 + 4})$, contain a positive real part, which implies instability and local growth of motion (i.e., a hyperbolic function of time). Throughout this work we will be primarily interested in the lightly damped case where the CEs contain both real (generally negative) and imaginary parts, signifying oscillatory behavior.

7.2 An Approximate Treatment

Let's again consider the behavior of the Hamiltonian system (with $\beta = 0$). We have already seen how the phase trajectories are contours of total constant energy. We have also seen how, for small-amplitude motion, we can obtain an analytic expression for the time evolution (simple harmonic motion). Ideally, we now wish to obtain the solutions in time for arbitrary initial conditions. Of course, it is a relatively simple matter to simulate the equation of motion using numerical integration, and in general, this will be the most appropriate method of attack. It should be noted that most integration routines are less liable to exhibit numerical problems for dissipative systems. However, we can also obtain reasonably good approximate answers using various analytic techniques. These have the advantage of providing a good degree of insight via closed-form solutions with explicit parameter dependence.

In this spirit, we can then consider the case in which we relax (somewhat) the stipulation that δ be very small by retaining terms up to quadratic in the expansion while still focusing on motion contained within a single potential energy well. In this case we see that an asymmetry is encountered, and the equation of motion for *moderately* large motion about the $x = -1$ equilibrium position (say) is now given by

$$\ddot{\delta} + 2\delta - 3\delta^2 = 0, \tag{7.10}$$

that is, the particle experiences a disproportionately weaker restoring force as it approaches the hilltop. However, we are now no longer in a position to obtain a *simple* closed-form exact analytic expression for the solution in time. Since we will examine solutions about the stable equilibrium, it will be convenient to make the coefficients unity (but remembering that the new x is shifted from the original x by 1). This was the rationale for scaling the complete equation of motion (5.38) to have the factor of 0.5 in the stiffness. The following form (one of the standard forms introduced in Chapter 4) can be considered as canonical and relates to the generic fold catastrophe:

$$\ddot{x} + x - x^2 = 0, \tag{7.11}$$

with static equilibria at $x_e = 0$ (stable) and $x_e = 1$ (unstable). This form clearly allows for unbounded solutions, which *escape*, certainly when $x > +1$.

We now make the assumption that the solution will not be far from harmonic and assume a truncated Fourier series solution of the form

$$x(t) = A_0 + a \cos(\omega t), \tag{7.12}$$

where the constant A_0 has been included because of the asymmetry of the restoring force. The implicit assumption has been made that the mass is given a nonzero initial position but with a zero initial velocity in order to eliminate the sine terms. Substituting this in Equation (7.11), we get

$$-a\omega^2 \cos(\omega t) + A_0 + a \cos(\omega t) - A_0^2 - 2A_0 a \cos(\omega t)$$
$$- a^2 \cos^2(\omega t) = 0. \tag{7.13}$$

Making use of trigonometric identities, ignoring the higher harmonics (the justification for this is not rigorous and some explanation can be found in (Jordan and Smith, 1977)), and setting the coefficients of constant and harmonic terms equal to zero gives

$$-(1/2)a^2 + A_0 - A_0^2 = 0,$$
$$-\omega^2 + 1 - 2A_0 = 0. \tag{7.14}$$

The second of these expressions tells us that the frequency of the assumed response depends on the offset and hence amplitude:

$$\omega = \sqrt{1 - 2A_0}. \tag{7.15}$$

The first expression tells us that the constant (i.e., the location of the center of the oscillation) is given by

$$A_0 = (1/2) \pm (1/2)\sqrt{1 - 2a^2}. \tag{7.16}$$

Substituting Equation (7.16) in Equation (7.15) gives

$$\omega = (1 - 2a^2)^{1/4}, \tag{7.17}$$

where we assume that $a^2 < 1/2$, and we see how the period of motion grows with amplitude (as expected) for this basically softening nonlinearity. For small-amplitude motion we have $a \to 0$, which implies a constant natural frequency of unity.

The general technique of matching harmonic terms in this way is referred to as the method of harmonic balance. Further details, including application to forced systems, can be found in Refs. (Hayashi, 1964) and (Jordan and Smith, 1977). This approach therefore has some nonlinear feedback into the solution, but it is only valid for moderate excursions about either equilibrium point, and certainly only for motion contained in the local potential energy well. Higher harmonics can be included; for example, a term $b \cos(2\omega t)$ can be added to the assumed solution but at an increasingly heavy algebraic cost. In fact, the resulting algebraic equations will then need to be solved numerically, thus immediately spoiling a prime advantage of having a closed-form solution. The justification for ignoring the higher harmonics is also parameter dependent in the general case. Some methods that are more consistent in a mathematical sense are based on perturbation theory and averaging (Nayfeh and Mook, 1978).

7.3 A Perturbation Approach

We have already mentioned how the underlying linear response can be used as a basis for expanding into the nonlinear regime. There are numerous techniques available. We shall use Lindstedt's method (Jordan and Smith, 1977; Nayfeh and Mook, 1978). We can apply this to Equation (7.11) by assuming that the quadratic effect is small, and we do this by placing ϵ, a small parameter, in front of this term. In contrast to the regular perturbation method, Lindstedt's method also expands the frequency as an unknown. Since we know that the frequency of oscillation reduces to the linear natural frequency $\omega_0 = 1$ for small-amplitude motion, we write

$$\omega = 1 + \epsilon\omega_1 + \cdots \tag{7.18}$$

and expand the solution as

$$x(\epsilon, t) = x_0(t) + \epsilon x_1(t) + \cdots . \tag{7.19}$$

As discussed in Ref. (Jordan and Smith, 1977), it is algebraically convenient to rearrange this by rescaling time using $\omega t = \tau$ so that (7.11) becomes

$$\omega^2 x'' + x - \epsilon x^2 = 0, \tag{7.20}$$

where a prime now denotes differentiation with respect to the scaled time, τ. We thus seek solutions with period 2π in τ. Substituting Equations (7.18) and (7.19) into Equation (7.20) gives

$$(1 + 2\epsilon\omega_1 + \cdots)(x_0'' + \epsilon x_1'' + \cdots) + (x_0 + \epsilon x_1 + \cdots)$$

$$- \epsilon\left(x_0^2 + 2\epsilon x_0 x_1 + \cdots\right) = 0, \tag{7.21}$$

and assembling like powers of ϵ (and setting their coefficients equal to zero) we have

$$x_0'' + x_0 = 0 \tag{7.22}$$

for ϵ^0, and

$$x_1'' + x_1 = x_0^2 - 2\omega_1 x_0'' \qquad (7.23)$$

for ϵ^1, and so on.

At this point (and only for autonomous systems) it is convenient to restrict initial conditions to

$$x_0(0) = a \qquad (7.24)$$

with all other initial conditions set equal to zero. Now we can write the solution to Equation (7.22) (the linearized problem) as

$$x_0(\tau) = a \cos \tau, \qquad (7.25)$$

and placing this in Equation (7.23) results in

$$x_1'' + x_1 = (a^2/2)(1 + \cos 2\tau) + 2\omega_1 a \cos \tau. \qquad (7.26)$$

This expression only allows periodic solutions if $\omega_1 = 0$, since the last term in the above expression (a secular term) would lead to growing unstable motion (i.e., an unbounded resonance). Hence we have the solution

$$x_1(\tau) = a^2/2 - a^2/3 \cos \tau - a^2/6 \cos 2\tau. \qquad (7.27)$$

Therefore, at this stage, the complete solution is

$$x(\tau) = a \cos \tau + \epsilon a^2/2 - \epsilon a^2/3 \cos \tau - \epsilon a^2/6 \cos 2\tau, \qquad (7.28)$$

and in terms of the original time scale it is

$$x(t) = a \cos \omega t - \epsilon a^2(2 \cos \omega t + \cos 2\omega t - 3)/6, \qquad (7.29)$$

where $\omega = 1$ to the first order in ϵ.

It is instructive to compare this solution (7.29) with the expression obtained from the method of harmonic balance, Equation (7.12), by considering how they deviate from the linearized solution on the one hand and the "exact" (numerical) solution on the other. We can retain more terms in the expansion (e.g., include terms of $O(\epsilon^2)$) but at the expense of increased algebraic complexity in much the same way as retaining more terms of the Fourier series in the method of harmonic balance (Bogoliubov and Mitropolsky, 1961). Figure 7.2 shows a comparison among the harmonic balance and Lindstedt solutions based on Equations

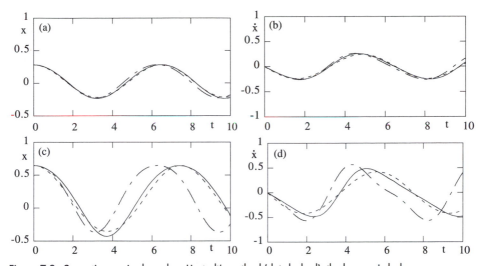

Figure 7.2: Some time series based on Linstedt's method (dot-dashed); the harmonic balance method (dashed); and the exact solution (continuous). $\dot{x}(0) = 0$; (a) and (b) $x(0) = 0.282$, (c) and (d) $x(0) = 0.646$.

(7.12) and (7.29) and the exact solution plotted as position and velocity time series. Time series generated using smaller initial conditions were effectively indistinguishable for the various solutions. We can thus envision these solutions existing in the vicinity of the negative energy well in Figure 7.1.

We note the second harmonic appearing in the Lindstedt solution, which was neglected in the harmonic balance solution (if retained there, a numerical procedure would have been required to evaluate the coefficients). The natural period for the harmonic balance that results, reflecting the softening spring effect, is very close to the exact natural period. For the harmonic balance solution we note that the ambiguity of ignoring higher harmonics is perhaps more justified in the more commonly studied example of a cubic stiffness (e.g., the moderate angle approximation to the pendulum (Baker and Gollub, 1996)). We also mention that for oscillators with a (symmetric) cubic restoring force Lindstedt's method does give an amplitude-dependent natural frequency, rather than the constant value (2π) obtained for this asymmetric oscillator, using this order of solution. Another perturbation method, the method of multiple scales, provides a powerful alternative to obtaining approximate analytic solutions (Nayfeh and Mook, 1978).

7.4 Phase Trajectories

If no restriction is placed on the initial conditions and a solution in time is required (rather than phase trajectories), it is possible to extract the time evolution by performing an integration over the closed path. Alternatively, a solution in terms of elliptic integrals (for undamped systems) can be obtained, but in general we use numerical integration. However, one of the real utilities of approximate analytical techniques will be made apparent for obtaining the periodic (both stable and unstable) solutions of forced, damped oscillations.

Before moving on to look at free oscillation data from the experimental system, we will briefly consider the accuracy, or range, of the solutions obtained by the harmonic balance and Lindstedt methods in terms of their phase planes. We can accomplish this quite easily for unforced, undamped oscillations since exact analytical solutions are readily available in the phase plane as contours of total constant energy.

We can separate variables as before to obtain the general expression for phase trajectories:

$$\dot{x} = \pm\sqrt{C - x^2 + (2/3)x^3}. \tag{7.30}$$

The value of C is determined by the initial conditions. We immediately note that C will have a strong effect on the qualitative nature of the curves: For two values of C (0 and 1/3) we have the equilibrium points; for small C (and small x) we have essentially linear motion giving ellipses, since the particle behaves as if it is oscillating within a parabolic potential energy well. The phase trajectories are bounded only over a certain range of values. The special phase trajectory that separates bounded from unbounded motion in this case is very similar to one side of the separatrix in the original double-well system. We shall see the crucial role this *homoclinic* solution plays in the global behavior of forced nonlinear oscillators.

For the moderate-amplitude oscillations from the previous section we can again make comparisons with the exact solutions (Equation 7.30 with the appropriate initial conditions, i.e., $x(0) = 0.282$, and hence $C = 0.0646$, and $x(0) = 0.646$, and hence $C = 0.2376$). This is shown in Figure 7.3 for relatively large oscillations, and these results correspond to the same solutions as those in Figure 7.2. This suggests that both sets of results are quite good (at these amplitudes) although the

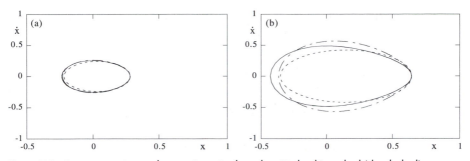

Figure 7.3: Some approximate phase trajectories based on Lindstedt's method (dot-dashed) and the harmonic balance method (dashed); the exact solution (continuous). $\dot{x}(0) = 0$; (a) $x(0) = 0.282$, (b) $x(0) = 0.646$.

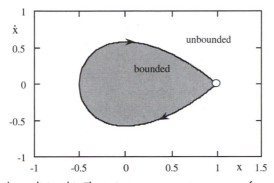

Figure 7.4: The homoclinic orbit. The trajectory represents a contour of constant total energy passing through the potential hilltop (unstable equilibrium).

asymmetric waveform of the Lindstedt results (pointing toward $x = 1$) is more appropriate, as we shall see.

Again, the homoclinic solution is easily determined by placing $C = 1/3$ in the expression for contours of constant total energy (Equation 7.30), which is shown graphically in Figure 7.4 for $x < 1$. Note the similarity with one side of the double-well energy contours (including the shift in origin). We can also write the solution in (positive and negative) time as

$$x(t) = 1 - 3/(1 + \cosh t), \qquad (7.31)$$

which again is very similar in form to the double-well homoclinic orbit given by Equation (7.4). This particular solution will be revisited later, as it is one of the building blocks upon which Melnikov theory is based in order to predict the onset of fractal basin boundaries in the forced vibration problem. However, for this Hamiltonian system, we can imagine the particle being nudged (infinitesimally) from the hilltop (at $t = -\infty$), moving down the valley and passing quickly through the minimum, changing direction at the same level of potential energy as the hilltop (at $x = -0.5$), and eventually moving back to the hilltop (at $t = +\infty$). Furthermore, any trajectory initiated within the shaded region will result in "bounded" periodic motion. Later we will see that for typical forced, damped oscillators the watershed between bounded and unbounded motion is an extremely complex issue.

It must be reiterated at this point that the canonical $(x - x^2)$ stiffness is based on a quadratic fit to the (cubic) restoring force, that is, the coefficients were set to unity in Equation (7.10) and the new origin shifted to the stable equilibrium position. The asymmetric and symmetric equations are almost the same over a certain range of x (see Chapter 4).

For intermediate values of total energy, we therefore expect asymmetric (egg-shaped) orbits that are slightly offset from the origin; this is where the approximate analytic solutions are appropriate (see Figure 7.3).

7.5 Experimental Free Oscillations

Initially, we will confine ourselves to the consideration of relatively small oscillations from the experimental perspective where the cart experiences motion confined within a small distance of a stable equilibrium in a similar manner to that described above. We have already touched upon this type of behavior in the experimental determination of the parameters of the system.

Figure 7.5 shows (a) a phase trajectory and (b) time series of a typical free decay started from rest a small distance (0.23 track units) from equilibrium. The decay reflects the shrinking ellipse as a small amount of energy is dissipated during each cycle. We also observe that the decay is neither strictly linear nor exponential, thus suggesting the simultaneous action of both Coulomb and viscous damping, and close inspection of the final rest position actually indicates a very small

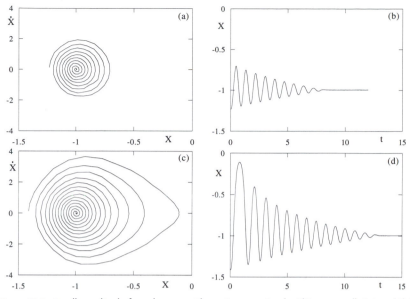

Figure 7.5: Small-amplitude free decays, with motion contained within one well. (a) and (b) $X(0) = -1.23$; (c) and (d) $X(0) = -1.41$.

discrepancy from -1 owing to sticking. The damped natural period in this case is 0.84 s, corresponding to a damped natural frequency of 1.15 Hz.

When the initial conditions contain a greater amount of mechanical energy ($X(0) = -1.41$), the effect of the nonlinear spring restoring force becomes noticeable (see Figures 7.5(c) and (d)). The softening spring effect experienced by the (initially) large-amplitude swings of a pendulum is a familiar example of amplitude-dependent periodic motion (Baker and Gollub, 1996). In the case considered here, we observe that the spring stiffness is asymmetric but can still be thought of as softening in the sense that moderately large motion is characterized by a natural period that is somewhat greater than that for the small-amplitude linear response. There is a noticeable distortion of the phase trajectory as it approaches the hilltop (at $X = 0$), and considerable lengthening of the first period of motion occurs. Although damping causes the gradual decay to zero passing through a regime of *linear* behavior, the approximate period–amplitude relation anticipated by Equation (7.17) is apparent, and a slight offset of the center of the oscillation can also be detected

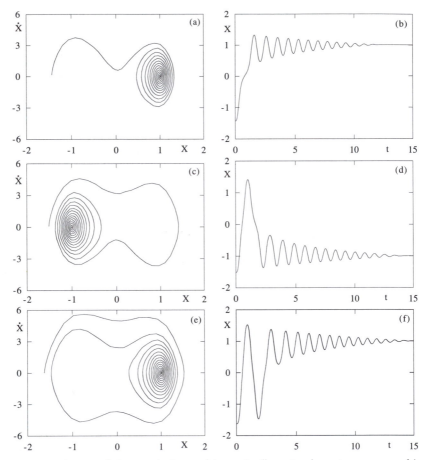

Figure 7.6: A series of phase projections and time series illustrating the various traverses of the hilltop as a function of initial position. (a) and (b) $X(0) = -1.43$; (c) and (d) $X(0) = -1.51$; (e) and (f) $X(0) = -1.62$.

for the first couple of cycles. The response remains confined within the left-hand well.

Suppose we start the cart with slightly greater initial energy, say with the initial conditions $(X(0), \dot{X}(0)) = (-1.43, 0)$. Figures 7.6(a) and (b) show the ensuing motion, again measured directly from the experimental apparatus. The slowdown near the origin is nicely displayed in the phase plane. Sufficient initial energy has been imparted to the system to cause the trajectory to traverse the hilltop (at $X = 0$) and enter the adjacent

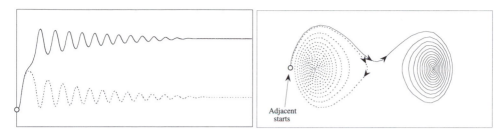

Figure 7.7: Two trajectories initiated on opposite sides of the separatrix.

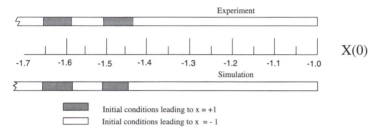

Figure 7.8: A one-dimensional slice through the initial conditions, starting from rest ($\dot{X}(0)$ = 0) at various initial positions.

potential energy well. The cart finally comes to rest at the $X = +1$ equilibrium. On further increase in the distance of the initial starting position from the bottom of the right-hand well, we observe multiple hilltop traverses. At this point it is interesting to note that experimentally it tends to be much easier to prescribe the initial position rather than the initial velocity.

We can superimpose adjacent starting positions to highlight the dependence on initial conditions. Figure 7.7 shows how these trajectories initially straddle the separatrix. This aspect is also somewhat related to the way in which an inverted pendulum will rotate in either direction given an uncertainty in the magnitude and direction of a small disturbance. However, the watershed dividing the trajectories in Figure 7.7 is a stable manifold, albeit one associated with an unstable equilibrium, and hence it can be located (in principle) for special combinations of initial position *and* velocity. As mentioned previously, in an experimental setting such as the one under consideration, the only readily accessible initial velocity is the rest state, although a powerful stochastic technique

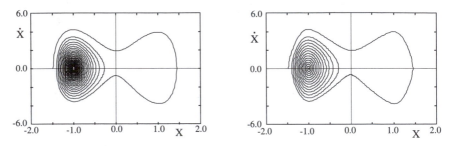

Figure 7.9: Comparison of numerical simulation based on (a) Duffing's equation (7.1) and (b) the equivalent result using the full equation (5.38).

will be introduced toward the end of the book. We can view the results shown in Figure 7.6 as a traverse along the position axis in the initial condition plane as shown in Figure 7.8.

We can also make a further assessment of modeling in the free decay. Figure 7.9 shows a comparison between numerical simulations for Duffing's equation (7.1) and the full equation of motion (5.38) without forcing. At least in this single case we see that Coulomb damping appears to provide a compelling correspondence with experimental results, especially as the motion amplitude becomes small. We also note in passing that the full equation of motion on which the simulations are based (Equation 5.38) gives phase trajectories that are somewhat more angular than the more familiar rounded curves of the standard Duffing equation (7.1).

Chapter 8

Forced Response: Periodic

8.1 Introduction

The forced response of the Duffing system in its three-dimensional phase space allows for the full spectrum of nonlinear behavior and serves to illustrate the often unpredictable nature of dynamics. We will restrict ourselves to harmonic excitation and start by considering the simplest case in which the magnitude of excitation is sufficiently small that the system responds in an essentially linear manner. Akin to the unforced case, when the nonlinearity becomes significant, many of the standard (predictable) features are lost. This is particularly the case with a three-dimensional phase space, and we will see that chaotic attractors and fascinating global behavior will join the more standard periodic attractors and local bifurcations. This plethora of complex behavior, of course, is what makes nonlinear systems so interesting.

This chapter will be introduced by looking at typical solutions from the forced version of Duffing's equation with viscous damping:

$$\ddot{x} + 2\zeta\dot{x} + Ax + Bx^3 = F\sin(\omega t + \phi). \tag{8.1}$$

We again consider initially the canonical form

$$\ddot{x} + \beta\dot{x} + x - x^2 = F\sin(\omega t), \tag{8.2}$$

where $\beta = 2\zeta$, and the stable point of equilibrium occurs at the origin (as opposed to the pair $\pm(A/B)^{1/2}$ from Equation 8.1 for negative A and positive B). We also note that the experimental system contains an explicit frequency-squared factor in the forcing amplitude that is not included here but can be considered as included in this definition of F (and where the previously used subscript on F has been dropped for convenience). Equation (8.2) is by no means easy to solve in the general case. However, we clearly expect to see an approximately linear response for small-amplitude motion, analogous to the autonomous systems of the previous chapter.

8.2 Mildly Nonlinear Behavior

Suppose we now seek to obtain *moderately* large-amplitude solutions of Equation (8.2) for typical lightly damped systems (e.g., $\beta = 0.1$). We will now make use of a simple approximate analytical technique to obtain solutions for various damping levels and forcing frequencies. The question of initial conditions and global issues will be dealt with in more detail at a later stage. Since the system will be undergoing oscillations in the asymmetric potential energy well, we might expect the unforced, undamped response to give us some clues about the form of moderately large-amplitude forced behavior.

Based on our knowledge of linear oscillators, it is natural to try a steady-state solution of the form of the first part of Equation (2.22) in a standard harmonic balance approach (Hayashi, 1964). However, we again add a constant term to allow for the expected asymmetry in the response, and we use both sine and cosine terms to allow for a phase difference (not necessary for the unforced problem):

$$x(t) = A_0 + a\cos(\omega t) + b\sin(\omega t). \tag{8.3}$$

Following the standard procedure (truncating and balancing terms), we arrive at the three coupled algebraic equations:

$$-A_0^2 + A_0 - (a^2/2) - (b^2/2) = 0,$$
$$-b\omega^2 - a\omega\beta + b - 2A_0 b = F, \tag{8.4}$$
$$-a\omega^2 + b\omega\beta + a - 2A_0 a = 0,$$

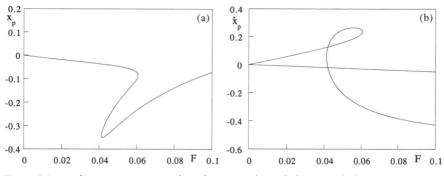

Figure 8.1: Analytic Poincaré section based on a single-mode harmonic balance solution. (a) Position versus forcing amplitude occuring at a fixed forcing phase; (b) corresponding velocity. $\omega = 0.85; \beta = 0.1$.

which can be combined to give

$$F^2 = 2\left(A_0 - A_0^2\right)\left[(\omega^2 + 2A_0 - 1)^2 + (\beta\omega)^2\right]. \qquad (8.5)$$

Again we are primarily interested in how the response amplitude changes with forcing parameters. We initially note that for $F = \beta = 0$, as expected, we recover the unforced response of Equation (7.15). Here ω is the effective frequency ratio (since the natural frequency has been set at unity); it is natural to assume that the output and input have the same frequency. We can envision the natural period versus amplitude (lengthening) relation as a *backbone* curve in the amplitude response diagram.

For small F, insufficient nonlinearity is induced to make the steady-state response significantly different from the first part of Equation (2.22). But let's now consider moderately large-amplitude motion and make comparisons with the experimental system. Equation (8.5) has one or three roots depending on the parameters. We will focus on frequency response behavior a little later, but first we make use of a Poincaré section by sampling the solution at a specific forcing phase. This is achieved by placing $t = 0$ in Equation (8.3). Hence the periodic attractor is now given as $x_p = A_0 + a$ and $\dot{x}_p = b\omega$. In this way we are taking an alternative view to the traditional amplitude and phase response. A typical result is shown in Figure 8.1, where the response is shown prior to any subsequent stability questions. The other (nearly constant) solution emanating from zero velocity in Figure 8.1(b) is actually associated

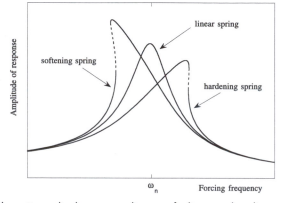

Figure 8.2: Schematic amplitude response diagrams for linear and nonlinear spring restoring forces. Dashed lines indicate an unstable branch.

with the (unstable) hilltop solution located at $x_p = 1$. We observe a resonance phenomenon analogous to the linear response shown in (2.6), but with some important differences. The softening spring effect causes a bending over of the curve and thus the appearance of coexisting solutions. This bending over of the response curves is actually toward lower frequencies in the conventional amplitude response diagram (i.e., a frequency sweep with a fixed forcing magnitude) and will occur in the opposite direction for hardening spring-type systems (for example for positive A and B in Equation 8.1) as shown schematically in Figure 8.2 (Holmes and Rand, 1976; Jordan and Smith, 1977; Collinge and Ockendon, 1979; Guckenheimer and Holmes, 1983; Schmidt and Tondl, 1986; Robinson, 1989; Worden, 1996). We will return to a hardening spring oscillator in a later chapter.

For relatively heavy damping we expect to see the single-valued case. However, under light damping (and moderately large forcing magnitude), the three roots for A_0 in Equation (8.5) are real, and a region of hysteresis is encountered. Thus, under slowly changing forcing frequency, a jump up or down occurs (depending on the direction of sweep). These jumps occur via the cyclic analogue of the saddle-node bifurcation and are associated with a CM of the system penetrating the unit circle in the complex plane at $+1$. In fact, this is one of the three generic mechanisms of instability under the action of a single control parameter as pointed out in the Chapter 3 (see Figure 3.6) (Arnold, 1988).

The transient part of the response can also be obtained by allowing the coefficients a and b to be functions of time in Equation (8.3). Following the standard harmonic balance procedure, two coupled first-order ordinary differential equations will result. The previously obtained steady states are now augmented by transients started from various initial a and b values and evolve in the plane of (a, b), the van der Pol plane (Thompson and Stewart, 1986).

This semianalytic procedure thus provides a degree of stability information. We also observe that the saddle-nodes that limit the region of hysteresis are associated with a vertical tangency in the response. However, the stability of the steady-state solutions must in general be determined using some kind of additional analysis based on behavior of small perturbations (Nayfeh and Balachandran, 1995). In the most general case, Floquet theory can be used to solve the resulting variational equation; that is, we can add a small perturbation to the steady-state solution:

$$x = x_0 + \eta. \tag{8.6}$$

Placing this in the equation of motion (8.2), we obtain

$$\ddot{\eta} + \beta\dot{\eta} + \eta(1 - 2X_0) = 0. \tag{8.7}$$

After substituting in the steady-state solution, where x_0 is given by Equation (8.3), we obtain the standard form of Mathieu's equation:

$$\ddot{\xi} + (\delta + \epsilon \cos t)\xi = 0, \tag{8.8}$$

which is a special form of Hill's equation (Hayashi, 1964). Transitions to instability resulting from solving Mathieu's equation thus relate to the generic routes to instability illustrated in Figure 3.6. The plane of (δ, ϵ), which depends on the parameters in the original equation of motion, thus separates into regions of local growth and local decay of perturbations. The typical loss of stability of period-one oscillations (i.e., oscillations with the same period as the forcing function) occurs either at the vertical tangency (fold or saddle-node bifurcation) or where the solution becomes unstable at the birth of a new period-two oscillation (flip bifurcation). These mechanisms were introduced in Figure 3.6.

In terms of the forcing parameters, it is insightful to plot the fold and flip bifurcations projected in parameter space. This is shown in

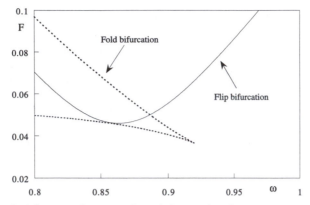

Figure 8.3: The bifurcations from period-one behavior plotted as a projection in the forcing parameter space. $\beta = 0.1$.

Figure 8.3, where a two-term harmonic balance (for improved accuracy) was used to obtain the steady-state information, followed by an application of Floquet theory (Nayfeh and Mook, 1978) to determine where the transition to instability occurred. The Routh–Hurwitz criterion is an alternative, and often simpler, approach to extracting the stability characteristics of a periodic orbit (Hayashi, 1964). The fold lines (dashed) are associated with the vertical tangencies in the response and form a cusp. The vertical tangencies in the response in Figure 8.1 for $\omega = 0.85$ can therefore be located in Figure 8.3, although this correspondence is not exact since the results in Figure 8.1 were obtained using a single harmonic whereas the results in Figure 8.3 were based on two harmonics. The flip line causes the start of a sequence of period-doubling bifurcations culminating in chaos. We note that this behavior is not apparent in Figure 8.1 because the instability results in a period-two oscillation, which bifurcates away from the now unstable period-one solution. These bifurcations have strong global implications, and their experimental verification will be left to Chapter 14.

Alternatively, the perturbation method (Nayfeh and Mook, 1978; Kevorkian and Cole, 1981) may also be used to obtain approximate solutions to the forced response. In general, there are a large variety of approximate analytical techniques for solving nonlinear ordinary differential equations (Jordan and Smith, 1977; Nayfeh and Mook, 1978).

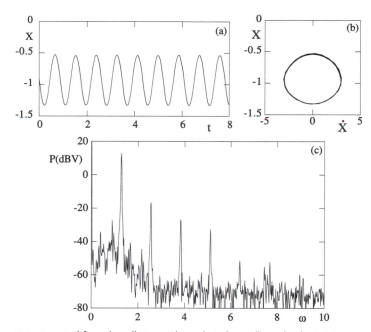

Figure 8.4: A typical forced oscillation with a relatively small amplitude. (a) Time series; (b) phase projection; (c) power spectrum. $\Omega = 1.0$; $F = 0.148$.

8.3 Simple Periodic Motion – Experiments

So far we have seen how, under the application of a time-periodic force, the typical long-term recurrent behavior is dominated by periodic, rather than point, attractors. Early experimental forced oscillation studies were conducted by Helmholtz, Rayleigh, and Duffing (Abraham and Shaw, 1982). We will consider forced oscillations initially in the vicinity of the underlying point attractor represented by the stable equilibrium at the potential energy minima ($X = \pm 1$) and gradually increase the energy of the external excitation to promote nonlinear behavior analogous to the preceding theory.

Suppose a sinusoidal base displacement is imparted to the system with a frequency of 1.15 Hz (i.e., near resonance) and a nondimensional magnitude of 0.148 track units. Recall that with this type of base excitation, we are effectively adding a frequency-squared term to the right-hand side of Equation (8.1). Figure 8.4 shows the resulting motion of the cart. In part (a), we see, as expected, that the frequency of the response is the same as the frequency of the excitation. The motion appears to be sinusoidal. The velocity is extracted from the time–position data using a central difference scheme. The position and velocity can then be projected onto the plane as shown in part (b). Note that here the position and velocity are interchanged from the usual convention for phase projections so that the position can be compared directly between parts (a) and (b). We can observe that the minimum amplitude is about -0.55 and the maximum is about -1.3; hence there is a slight spring softening effect at these parameters (i.e., the phase projection is not quite elliptical or centered at -1). Nonetheless, the periodic nature of the response is confirmed in the power spectrum of part (c). The first harmonic provides the dominant energy in the response, and the frequency scale is left in hertz (and related to Ω by $\omega_n = 1.15$). A Hanning window was used, and the FFT was based on the position time series.

A summary of these types of periodic attractor and their amplitude dependence on forcing frequency is shown in Figure 8.5 for data taken directly from the experiment. The jumps in amplitude span the region of hysteresis and thus verify the results of the previous section and confirm the dependence on initial conditions (but not the extreme sensitivity associated with fractal basin boundaries to be encountered in Chapter 14). For lower values of the forcing amplitude, a single-valued response similar to the linear response of Figure 2.6(b) is obtained, and we also note that the response approaches zero for low frequencies (the harmonic balance results were obtained for direct mass excitation).

It is possible to obtain cross-well periodic solutions from Duffing's equation (Jordan and Smith, 1977), but these occur for very large forcing conditions. A typical example will be illustrated in the final section of this chapter.

8.4 Subharmonic Behavior

The response shown in the previous section is quite close to primary resonance. Another commonly encountered behavior in nonlinear systems is a subharmonic, that is, a period-n oscillation that takes n forcing periods to complete a full cycle. This may also occur over a wide range of parameter values, and a typical (cross-well) output of a numerical

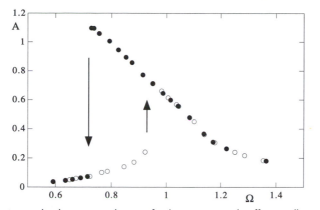

Figure 8.5: An amplitude response diagram for the experimental Duffing oscillator. Amplitude A is peak-to-peak. $F = 0.047$.

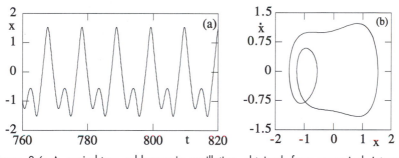

Figure 8.6: A period-two subharmonic oscillation obtained from numerical integration of Equation (8.1). (a) Time series; (b) phase projection ($\zeta = 0.15$, $A = -1$, $B = 1$, $F = 0.65$, $\omega = 1.2$, $\phi = 0$).

simulation of Equation (8.1) is shown in Figure 8.6 as a time series and phase projection after transients have been allowed to decay. This subharmonic oscillation is a feature not dissimilar to the response of an undamped linear oscillator forced at twice its natural frequency. This type of behavior can be analyzed using the previously developed approximate techniques. The assumed form of the solution can be made specific to the form of the response that is sought. For example, a harmonic balance solution can be based on

$$x(t) = A_0 + a_1 \cos(\omega t) + b_1 \sin(\omega t) + a_2 \cos(2\omega t) + b_2 \sin(2\omega t),$$
$$(8.9)$$

and a typical amplitude response (including unstable branches) is shown

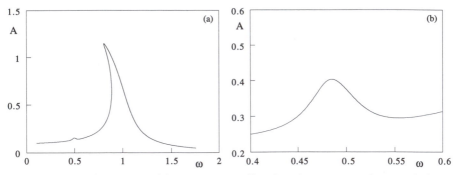

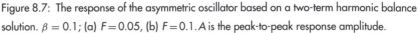

Figure 8.7: The response of the asymmetric oscillator based on a two-term harmonic balance solution. $\beta = 0.1$; (a) $F = 0.05$, (b) $F = 0.1$. A is the peak-to-peak response amplitude.

in Figure 8.7. This is the more conventional analogue to the Poincaré sampled response of Figure 8.1, and here the peak-to-peak amplitude (chosen because of the asymmetry) is used as the response measure. Part (a) shows the appearance of a slight bump in the vicinity of $\omega = 0.5$. This is the beginning of a subharmonic resonance. Part (b) shows a close-up view when the forcing magnitude has been doubled to $F = 0.1$. A time series at this frequency shows a significant second harmonic component. It is also interesting to note that for significantly higher forcing, this sub-harmonic peak starts to bend over, and corresponding instabilities can occur (i.e., the bifurcations from Figure 8.3 are repeated in a similar fashion for various lower frequencies (Thompson, 1989)). However, in general, the solution is a Fourier series, and as many harmonics can be included as desired, given the constraint of rapidly diminishing algebraic tractability. The occurrence of a subharmonic of order two is also associated with the direct loss of stability of a period-one branch via the flip bifurcation, as described earlier.

A typical $n = 2$ experimental response is shown in Figure 8.8. Here the forcing parameters have been changed slightly from those of Figure 8.4. The forcing amplitude is the same, but the forcing frequency has been lowered to $\Omega = 0.84$. We now observe a splitting of the amplitude to an alternating low/high response, which is manifest most clearly as the cart approaches the hilltop (at $X = 0$). The transition from the period-one to the period-two oscillation results from the generic period-doubling bifurcation detailed earlier. The additional frequency contribution at half the fundamental can be seen in part (c).

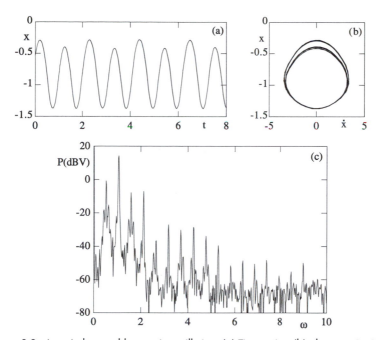

Figure 8.8: A period-two subharmonic oscillation. (a) Time series; (b) phase projection; (c) power spectrum. $\Omega = 0.84$; $F = 0.148$.

Equivalent period-one and period-two attractors located in the right-hand well were also present corresponding to almost mirror image attractors (notwithstanding a slight asymmetry in the overall track shape). Upon further reduction of the forcing frequency, a period-four oscillation appears, followed by a period-eight oscillation, and so on (Schmidt, 1986; Szemplinska-Stupnicka, 1988; van Dooren, 1988). This is the start of a sequence leading to chaos and will be investigated in more detail later. Superharmonics (i.e., oscillations that repeat themselves n times within a complete forcing cycle) may also appear in nonlinear oscillators (Nayfeh and Mook, 1978). The case of subharmonic forcing will be deferred until Chapter 13.

8.5 Nonstationarity

Figure 8.5 was obtained by fixing the forcing frequency at one of the extremes, starting the system initially from rest, allowing transients to decay, and plotting the resulting steady-state amplitude. The frequency was then incremented very slightly, and after a short settling period, the new amplitude was noted, and so on. In this way, transients were minimized since the step changes in the parameter were kept small, and the steady-state amplitude was followed until the basin of attraction (i.e., region of initial conditions that lead to a particular attractor) had disappeared. An alternative to this procedure is to reset the initial conditions to zero (say) after each increment. This approach was followed in one of the sections earlier, and the quiescent state obviously is a representative initial situation for a dynamical system. However, the former case does represent a relatively large class of problems in practice; the smooth (and slow) change in a system parameter is quite typical (Mitropolskii, 1965). In mechanical engineering, a rotating shaft cannot be made to instantaneously exhibit a desired rpm. Rather, there is a gradual *spin-up* and thus any resonance conditions that might be somewhat below the nominal operating speed will be encountered (at least for a short period), and similarly with *run-down*. This condition of an evolving parameter is generally called nonstationarity, and for a ramped (linearly varied) input it is how the resonant jump in amplitude is typically realized (Raman, Bajaj, and Davies, 1996; Thompson and Virgin, 1986). A related form of nonstationarity is the difficulty in holding all parameters of a problem fixed. Drift, for example, may occur because of changes of temperature. Nonstationary excitation is sometimes used as a tool for nondestructive testing (e.g., the swept sine wave (White, 1971)).

However, the nonstationarity of the frequency evolution is relatively easy to achieve experimentally by using a stepper motor on the Scotch-yoke mechanism to achieve a monotonic increase or decrease. In this way the transition through resonance can be followed continuously. Here we consider the nonstationary transition through the hysteretic jumps in amplitude for the twin-well experimental system. A similar study has been conducted to follow the period-doubling bifurcations that also occur in this system (Todd, Virgin, and Gottwald, 1996).

Referring back to the original track–cart model (5.38), we can incorporate a continuous change, r, in the forcing frequency in the following way. Taking the initial phase as zero, without loss of generality, and nondimensionalizing using $\tau = \omega_n t$, $X = x/x_e$, $R = r/\omega_n^2$, $\Omega = \omega/\omega_n$,

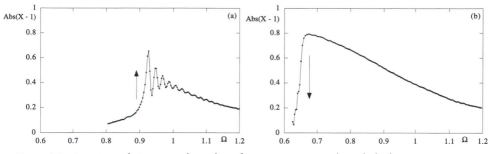

Figure 8.9: Experimental spin-up and run-down frequency responses through the hysteresis region. $F=0.05$; (a) $R=0.0005$, (b) $R=-0.0005$.

and $F = f/x_e$, we have for the right-hand side of Equation (5.38)

$$F\sqrt{(\Omega_0 + R\tau)^4 + R^2}\sin[\theta(\tau) - \phi(\tau)], \qquad (8.10)$$

where

$$\theta(\tau) = \Omega_0\tau + \frac{1}{2}R\tau^2,$$
$$\tan[\phi(\tau)] = (\Omega_0 + R\tau)^2/R, \qquad (8.11)$$

and where Ω_0 represents the starting frequency. From measurements made prior to this experiment, it was found that $\alpha^2 = 0.596$, and again acknowledging the difficulty of measuring damping, the values $\zeta \approx 0.002$ and $\mu \approx 0.02$ were estimated.

Figure 8.9 shows some typical results taken from the experiment. The numerical simulation results were very similar but are not included. Here, the amplitude of the motion is taken about the potential energy minimum, that is, $|X - 1|$. Part (a) shows a typical spin-up result. The amplitude of excitation is held fixed ($F = 0.05$), and for the fixed rate of frequency evolution ($R = +0.0005$), the jump *up* to the resonant branch of the solution is slightly delayed owing to the non-stationarity. Under stationary excitation at this level, we would expect the saddle-node to occur at frequencies slightly lower than $\Omega = 0.9$. The postjump oscillations are caused by a combination of the transient motion caused by the finite jump and the evolution through resonance: There is a sudden shift in phase lag between the input and output at resonance (including linear resonance) that takes time to adjust. Part (b) illustrates the corresponding run-down with a jump *down* to the

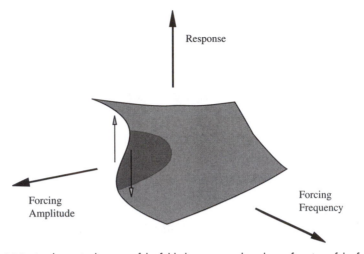

Figure 8.10: A schematic diagram of the folded response, plotted as a function of the forcing parameters.

nonresonant branch. In this case, the postjump behavior is dominated by the presence of Coulomb damping and, since the transmissible response tends to zero for low frequency ratios (see Figure 2.6(b)), the motion practically ceases.

Figure 8.9 was based on a typical (i.e., slow, but nonnegligible) rate of evolution, and the overshoot, for example, will be more or less dependent on this rate. In the extreme case, the evolution may be fast enough such that the resonance and jump can be effectively eliminated.

We recognize these evolutions as traverses along the folded response *surface* (see Figure 8.10) at a specific forcing amplitude. These resonant jumps occur at the vertical tangencies and mark the range of hysteresis, a classical feature in nonlinear resonance. For lower forcing levels the response is single-valued and not dissimilar to the linear resonance described in Chapter 2. This surface is familiarly known as the *cusp* catastrophe (Poston and Stewart, 1978), a name derived from the projection of the multivalued response regions down onto the parameter plane (e.g., producing the dashed curve in Figure 8.3). We shall see shortly how this response surface, although allowing for competing solutions, gives little hint of the extremely complicated behavior found under stronger forcing conditions.

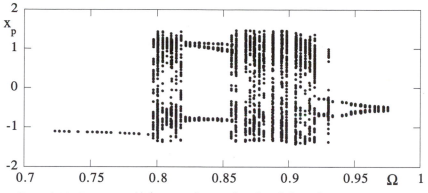

Figure 8.11: Experimental bifurcation diagram for relatively large forcing. $F = 0.205$.

8.6 Cross-Well Periodic Behavior

So far, we have discussed only periodic behavior relating to motion confined within the potential energy well, that is, only a mild distortion from a harmonic response. It is also possible to obtain a variety of cross-well periodic behavior, but generally the forcing conditions must be relatively strong. If the forcing parameters are increased further, we can again summarize the resulting behavior by plotting a bifurcation diagram. It is convenient to extract Poincaré points while sweeping (evolving in a slow nonstationary sense) through forcing frequency. A typical example is shown in Figure 8.11 for a forcing amplitude about four times that used to obtain the amplitude response diagram shown in Figure 8.5.

We observe that, over certain ranges of forcing frequency, the Poincaré points fill out a range of positions (taken at a specific forcing phase). We also point out that because this figure was obtained by slowly sweeping through the forcing frequency, minimal transients are induced, and thus, remote coexisting attractors (if they exist) are not necessarily revealed. The slight spread in the results toward the higher frequency end of the diagram is caused by proximity to the bifurcation (to the period-two solution) with some residual transient behavior. Also, the superposition of a reverse sweep (used in Figure 8.5 to uncover hysteresis), achieved by increasing frequency, is not undertaken. The uneven spread of frequencies is caused by finite precision in prescribing the frequency rather than in actually measuring it. A more thorough investigation of dependence

115

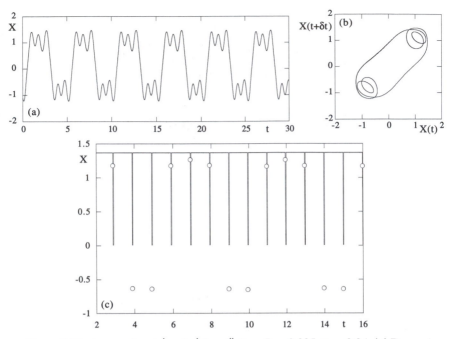

Figure 8.12: An experimental period-5 oscillation. $F = 0.205; \Omega = 0.84$, (a) Time series; (b) phase projection based on the position plotted against itself one quarter of a cycle later (δ based on a quarter cycle); (c) Poincaré sampling, which occurs at the spiked intervals.

on initial conditions is deferred until Chapter 14. The regimes of non-periodic behavior are where chaos appears, and this type of response will be scrutinized in the next chapter.

The periodic response pertaining to a frequency of $\Omega = 0.84$ is shown in Figure 8.12 as a time series in (a) and as a time-lag embedded reconstruction of the phase projection in (b). This is a very large amplitude (spanning both potential energy wells) oscillation, which takes five forcing periods before repeating itself. This is apparent from part (c), which shows how the time series is stroboscopically sampled once every forcing cycle to produce five repeating points. It is interesting to see that although there only appears to be three Poincaré points in Figure 8.11 at this frequency, this is just an illusion caused by the choice of (arbitrary but fixed) forcing phase and the *projection* of the Poincaré points as a function of frequency ratio. A period-ten oscillation was also observed for other forcing conditions; this appeared to be the consequence of a period-doubling bifurcation from a period-five motion.

Chapter 9

Forced Response: Nonperiodic

9.1 Introduction

So far, we have seen the appearance of essentially linear behavior for relatively small-amplitude forced oscillations together with the occurrence of (local) instability phenomena and subharmonic oscillations. The main features were a growth of amplitude with forcing magnitude and proximity to resonance and a response frequency the same as the input (forcing) frequency – features not dissimilar to the purely linear oscillator. The appearance of hysteresis signaled the enhanced role played by nonlinearity and initial conditions. However, considered locally from a topological viewpoint, all these responses can still be classified as periodic attractors, and in many important ways they are not substantially different from their point attractor counterparts in dissipative, gradient systems. Predicting the future behavior in these cases is relatively easy. But for nonlinear dynamical systems (flows) that exist in a phase space of three or more dimensions thoroughly more complicated and less predictable behavior becomes possible (Lorenz, 1963).

9.2 Chaos

It is the fascinating (and universal) nature of *chaos* that will be the main focus of attention in this chapter. The discussion will be somewhat constrained to the types of behavior exhibited directly by the experimental system, with a focus on invariant measures. A number of excellent books on chaos are available. A sample includes those covering theoretical (numerical) approaches (Guckenheimer and Holmes, 1983; Thompson and Stewart, 1986; Wiggins, 1990; Marek and Schreider, 1991; Ott, 1993), experimental aspects (Moon, 1992; Tufillaro Abbot, and Reilly, 1992), and general treatments (Jackson, 1989; Mullin, 1993). This subject has reached a sufficient level of maturity that there are even books using pedagogical approaches (Abraham and Shaw, 1982; Strogatz, 1994; Baker and Gollub, 1996) and more general expositions for the general public (Gleick, 1987; Stewart, 1989).

In keeping with the progression of the previous chapters we introduce further strengthening of the external driving, thus encouraging significant nonlinear effects. It will be shown later that a progression toward chaotic behavior follows some very generic routes, but since many of the subtle interactions are of a global nature, they will be left until later (Dowell and Pezeshki, 1986).

At this point, we simply illustrate some typical chaotic features to contrast this type of behavior with the (predictable) behavior already described. The possibility of extremely complicated dynamics was envisioned theoretically by Poincaré (Poincaré, 1921) and Birkhoff (Birkhoff, 1941), and the careful experimental work of Reynolds (Reynolds, 1883) and Rayleigh (Rayleigh, 1945) also gave some hint of complexity and certainly highlighted the limitations of linear theory. However, it was the advent of digital computation that allowed the realization that chaos was quite common, robust, and worthy of study (Lorenz, 1963). Although many of the early studies (and subsequent quantitative tools (Swinney, 1983; Broomhead and King, 1986)) on chaotic behavior were based on numerical investigation, it is the experimental verification of chaos that has placed the subject on a solid foundation in the context of engineering vibration from a nonlinear dynamics perspective. Many mechanical engineering systems are inherently nonlinear and increasingly so, given, for example, the quest for lighter, flexible components (using new materials) in increasingly hostile operating environments.

Figure 9.1 shows time series generated by subjecting the experimental system to a moderately strong base excitation. The top part of this figure shows the response after a considerable number of transient cycles have been allowed to decay. The system spends time oscillating in a

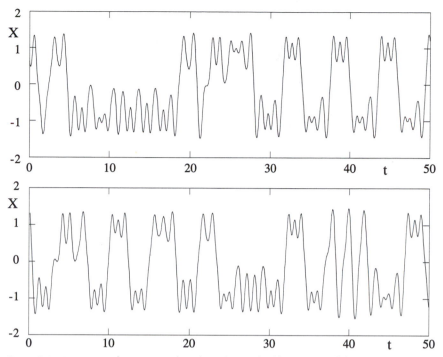

Figure 9.1: Experimental time series describing the randomlike position of the oscillator as a function of time. $\Omega = 0.84$; $F = 0.188$.

somewhat erratic manner about, and between, the two equilibria located at $X = \pm 1$. Although the motion never quite repeats itself, there are passages of nearly recurrent behavior, and the dominant frequency is clearly about 1 Hz (the forcing frequency). The nondimensional forcing magnitude of 0.188 units is obviously sufficient to allow interwell, spillover motion. It will be seen later that the transition between motion bounded within a single potential energy well and motion that escapes is interesting indeed.

The lower part of this figure shows another portion of the time series taken from the same experimental run but after an extended time lapse. Clearly, this motion does not appear to be "settling down". The corresponding phase projection is shown in Figure 9.2, and again, the influence of the underlying static equilibria is apparent. Conducting an FFT on the above output leads to the rather broadband power spectrum shown in Figure 9.3(a) with energy present over a wide range of frequencies (Bergé, Pomeau, and Vidal, 1984; Brunsden, Cortell, and Holmes,

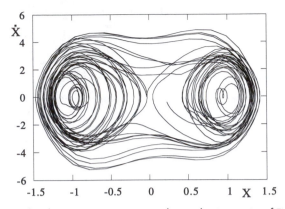

Figure 9.2: The phase projection corresponding to the time series of Figure 9.1.

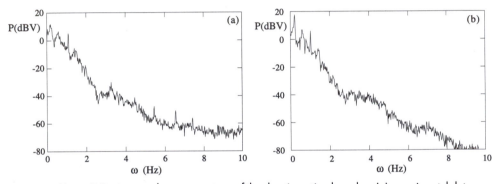

Figure 9.3: Averaged power spectrum of the chaotic motion based on (a) experimental data, (b) the numerical model.

1989). The power spectrum in Figure 9.3(b), which is numerical (based on Equation 5.38), continues to give an indication of the close correlation between the numerical and experimental models. In this instance, the broadband nature of the power spectrum is clearly not the result of noise or stochastic effects. In both of these cases, the data manipulation approaches were the same, and results were averaged over eight time series each.

9.3 Poincaré Sections

The stroboscopic sampling technique, based on a first return map introduced in Section 3.3, can be employed to great effect in shedding light on this kind of chaotic response. As mentioned earlier, the reduction from a flow in three dimensions to a map in two dimensions is particularly useful in this type of forced nonlinear oscillator. In the experiment, it is relatively easy to extract a data point at the appropriate phase once during each forcing cycle using a simple photodetector or magnetic trigger (both of which were used). This process thus achieves synchronizing the sampling with the forcing frequency (as described in Chapter 6). The analogous approach in numerical simulation is extremely straight-forward, whereby the time step is chosen such that an integer number of time steps (typically 50 or 100) occurs during a complete forcing cycle. Using this approach, we find the sampled response shown in Figure 9.4(a) based on experimental data, together with the equivalent numerical response in part (b).

Both of these mappings contain 10,000 points and part (a) corresponds to a real-time experimental run of about three hours. An alternative section can be constructed by shifting the phase at which the surface of section is made. The Poincaré section for this is shown in Figure 9.5 when the phase shift is $\pi/2$. A collage of these sections can be built up by sweeping through the forcing phase (Gottwald, Virgin, and Dowell, 1992).

The degree of fine structure is quite remarkable. Following the evolution of the Poincaré section as a function of the phase of sampling, the folding and stretching nature of the attractor become apparent (Gottwald, Virgin, and Dowell, 1992). It should be remembered that there is still an overall contraction of trajectories onto this *strange* attractor, but a local exponential divergence within the attractor also occurs. This results in an extreme sensitivity to initial conditions and the mixing of trajectories (Holmes, 1979; Holmes and Moon, 1983).

At this point it is also quite interesting to return to the modeling of Chapter 5 and compare this with the Poincaré section resulting from a simulation of the simple (transmissibly forced, viscously damped) Duffing equation (4.1), which the original experiment set out to mimic. Figure 9.6 shows this comparison. Here a higher damping value of $\zeta = 0.025$ was used to compensate for the absence of a Coulomb damping term. The basic features of the attractor shown in Figure 9.4 are captured, although certain minor differences are apparent. As a final comparison, a measured chaotic attractor from the magnetically buckled beam experiments of Moon (Moon, 1992) are reproduced in Figure 9.7.

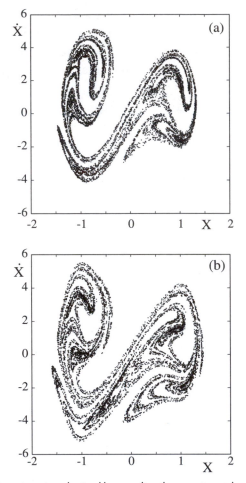

Figure 9.4: (a) Poincaré section obtained by sampling the experimental cart motion at forcing phase $\phi = 0$. (b) The corresponding Poincaré section from the simulation. $F = 0.188$; $\Omega = 0.84$.

Despite the continuous (and quite different) nature of this system the fine structure of the behavior is also apparent.

9.4 Autocorrelation

Another useful tool in the analysis of chaos is the autocorrelation function (Bayly et al., 1994). The basic idea here is to evaluate the

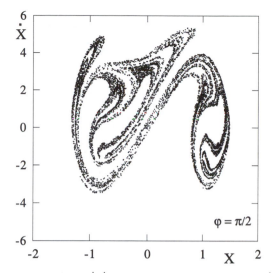

Figure 9.5: The same experimental chaotic response as in Figure 9.4(a) but sampled at a forcing phase of $\pi/2$.

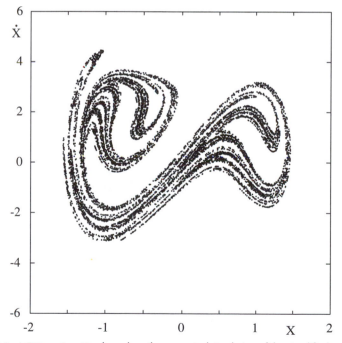

Figure 9.6: A Poincaré section based on the numerical simulation of the simplified numerical model.

123

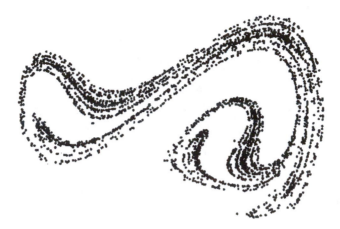

Figure 9.7: A Poincaré section based on the buckled beam experiments of Moon. Reproduced with permission from Moon (1992).

function

$$A(\tau) = \lim_{T \to \infty} \frac{1}{T} \int_0^T x(t)x(t + \tau)\,dt \qquad (9.1)$$

over a number of time scales (based on the time lag, τ), to establish (even small amounts of) repeatability in a signal. In practice this integral is replaced by an equivalent summation. From this equation, it is apparent that a constant signal will always be perfectly correlated with itself, a periodic signal will have a periodic correlation (as a function of the time delay τ), and white noise will have zero correlation for *any* later time. Applying this measure to the chaotic time series obtained from the cart experiment results in the autocorrelation function plotted in Figure 9.8. We note that the local maxima reflect a degree of underlying period-five and period-one recurrence. This can also be observed in the chaotic time series (see Figure 9.1). A number of periodic *windows* appear in the parameter regime of chaotic behavior including a stable period-five oscillation not dissimilar to the behavior found between 30 and 50 seconds in Figure 9.1. A stable period-five oscillation is shown in Figure 8.12. In fact, the appearance of an odd, low-order subharmonic is often a clue that chaos may be exhibited (Li and Yorke, 1975).

We also note at this point that the first zero crossing (i.e., where $A(\tau)$ drops to zero) of the autocorrelation function is often used as

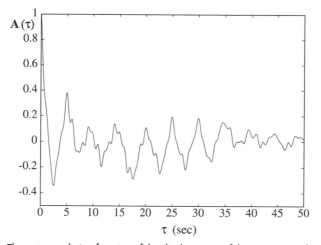

Figure 9.8: The autocorrelation function of the displacement of the experimental Duffing cart. Parameters are the same as in Figure 9.1. Local maxima in the autocorrelation function reflect a degree of recurrent behavior.

the appropriate time delay in phase space reconstruction methods using embedding (Casdagli et al., 1991), although in this case the presence of the dominant period-five behavior obscures this to a certain extent. The time-delay embedding approach will be adopted in later reporting of results because of the practical shortcomings of numerical differentiation mentioned earlier.

9.5 Dimension

The fine structure of the chaotic attractor (and specifically a degree of self-similarity at closer inspection) has the appearance of a fractal object (Mandelbrot, 1983). Although there are various definitions of fractals and dimension (Moon, 1992), it is apparent that fractals possess a relatively complex geometry, and it is the manner in which points fill a space that provides a measure of dimension (Grassberger and Procaccia, 1983; Farmer, Ott, and Yorke, 1983; Badii and Politi, 1985). We can think of a point as having a dimension of zero, a line as a one-dimensional object, and so on. A fractal object (e.g., the Cantor set or Koch curve) has a noninteger dimension. One of the most commonly used algorithms

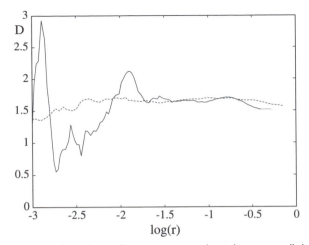

Figure 9.9: The curves of correlation dimension D versus $\log r$ showing a well-defined plateau near $D = 1.7 \pm 0.1$. Experiment: solid curve; simulation: dashed curve.

is based on the correlation dimension:

$$D = d[\log C(r)]/d[\log r], \qquad (9.2)$$

which is computed by constructing a circle (or sphere) around each point and counting the number of points in it and then extracting a power-law scaling as a function of radius r. In essence this relates to the probability of finding two points contained in the same area. This technique was applied to the data of Figures 9.6 and 9.4(a), and the result is shown in Figure 9.9. The value of $D \approx 1.7$ (corresponding to the plateau) is not unreasonable for a two-dimensional map. The effects of noise and a finite data set (providing limiting estimates for small and large regions in the experimental data) can also be seen in this figure. This computation is much less straightforward for high-order systems where a considerable amount of data is required to draw firm conclusions (see Appendix B). We also note that the dimension of an object in a Poincaré section is one less than the dimension of the corresponding flow. We shall see in the following section that when the Poincaré surface of section occurs at equal time intervals (e.g., when triggered by a specific phase of forcing) this change in dimension is related to the appearance of a zero Lyapunov exponent, an important invariant measure, which we turn to next.

9.6 Lyapunov Exponents

Finally, we introduce a key quantitative measure of chaos. We have already stated that initially adjacent trajectories diverge rapidly, and in fact, it can be shown that this divergence takes place, on the average and in a local sense, at an exponential rate. Mathematically, we can describe this process in its simplest form by

$$d(t) = d_0 e^{\lambda t}, \tag{9.3}$$

or alternatively

$$\lambda = \lim_{t \to \infty} (1/t)\ln[d(t)/d_0], \tag{9.4}$$

where d_0 is the initial (small) separation of points on the attractor, $d(t)$ is the distance between them at later times, and λ is the Lyapunov exponent (LE) (Eckmann and Ruelle, 1992; Kruel, Eisworth, and Schreider, 1993).

In the vicinity of a fixed point of a three-dimensional flow (as encountered in our forced oscillators), the three LEs are seen to be simple eigenvalues that scale the local evolution of the flow and whose sum is equal to the (volumetric) divergence of the flow (we still have the overall contraction of phase space even for the case of chaos). One of the LEs is zero, which is due to the fact that there is no stretching or contraction in the time direction. However, despite this overall contraction, an individual LE may be positive, which implies that some of the trajectories will locally diverge. This also implies a sensitivity to initial conditions. Hence, recalling that we are in general dealing with a bounded system, the existence of at least one positive LE is a criterion for chaos (Ott, 1993). The necessity of dealing with long-term average values is based on the fact that a short-lived positive LE may appear in the vicinity of a saddle-point, for example.

In the actual computation of LEs it is therefore only really necessary to compute the largest. However, calculating the full spectrum of exponents as a function of a control parameter clearly will contain a considerable amount of information (Brown, Bryant, and Abarbanel, 1990). If the equation of motion is known, then a fairly straightforward computation can be conducted based on the behavior of the variational equation near the fundamental (fudicial) trajectory. There are a number of computational issues, including the definition of *local*, locating

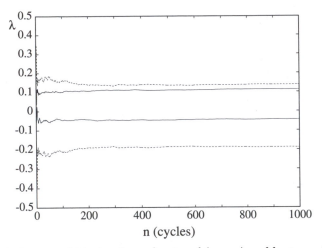

Figure 9.10: Estimates of LEs plotted as a function of the number of forcing cycles used in the average divergence calculation. Experimental estimates: $\lambda_1 = 0.11, \lambda_2 = -0.04$ (solid curves); simulation estimates: $\lambda_1 = 0.14, \lambda_2 = -0.19$ (dotted curves).

nearest neighbors, ill-conditioning (e.g., finding the appropriate time interval between successive QR decompositions (matrix factorizations) of the Jacobian map (Grebogi et al., 1984)), and the effect of finite data. This latter problem is especially significant for experimental data, which exhibit the various effects of noise.

For experimental data, it is possible to extract the largest LE using a technique based on successive linear approximations to the local flow (Wolf et al., 1985; Kruel, Eisworth, and Schreider, 1993). Other algorithms that can be applied to discrete data obtained from a Poincaré map have also been developed. An additional complication of extracting LEs from data is the determination of the Jacobian, but a local least-squares regression approach (Lathrop and Kostelich, 1989; Murphy et al., 1994) has been shown to work well and is described in more detail in Sections 9.7 and 11.4, where it is applied to unstable and stable fixed points, respectively. The technique of time-lag embedding (Section 6.4) also plays a role in establishing an appropriate space in which to study the dynamics (see Sections 6.5 and 13.3). Both of these approaches (using the equations of motion from a simulation, Figure 9.6, and from an experimental Poincaré map, Figure 9.4(a)) were applied to the chaotic responses and the results are shown in Figure 9.10.

It is important to recognize that these invariant measures are statistical in application, and hence, conclusions are subject to the same scrutiny as in any statistical study. These measures are often computed using a rather ad hoc approach, with the parameters varied and checks made. A useful relation (and hence check) between LEs and fractal dimension was discovered by Kaplan and Yorke (Kaplan and Yorke, 1979). They show that for a two-dimensional map of the type under study here, the dimension of the attractor can be estimated from

$$D_L = 1 - \frac{\lambda_1}{\lambda_2}, \tag{9.5}$$

where subscript 1 refers to the highest exponent. This expression is actually a specific case of the more general (Lyapunov dimension) relation

$$D_L = k + \frac{\sum_{m=1}^{k} \lambda_m}{-\lambda_{k+1}}, \tag{9.6}$$

where k is the maximum number of LEs that can be added together such that $\sum_{m=1}^{k} \lambda_m$ remains negative. This implies that the first $k + 1$ exponents are particularly influential. Applying this to the LEs computed from the simulation, we obtain $D_L = 1.737$, which is very close to the dimension estimate in Figure 9.9. However, using the above expression for the experimental data gives an estimated dimension of $D_L = 3.75$. It is likely that the discrepancy is caused by the strong effect of noise on the estimate of the second (negative) LE. The dimension of the original chaotic flow is obtained simply by adding 1 to the above values. The flow is clearly dominated by (and magnifies noise in) the expanding direction.

Another check on these computations can be made by referring back to the divergence theorem introduced in Chapter 3. The volumetric contraction of a small element of initial conditions was given in Equation (3.9). This time evolution is also governed by the LEs:

$$V(t) = V(0)e^{(\lambda_1 + \lambda_2 + \lambda_3)t}, \tag{9.7}$$

and we see that the divergence is also given by the sum of the LEs (equivalent to the CEs in the unforced linear oscillator). In Equation (9.7) the third LE, λ_3, is simply zero based on the time axis and is not computed when using the Poincaré map to extract the LEs. This sum, based on the numerical computation in Figure 9.10 and using the appropriate damping scaling (with $\omega_n = 1$), gives div(x_i) $= -0.05$, that is, -2ζ, twice the level of linear viscous damping used in this particular model (as expected from Equation 3.11). However, for the experimental results the sum of the LEs is 0.07, which is in error in both sign and magnitude. Again, the ill-conditioning of the algorithm places some doubt on the accuracy of the second LE. The computation of the LEs requires considerable data since they are based on long-term average properties. Furthermore, it is possible that Coulomb damping may play a complicating role for the experimental data. Alternatively, if the whole spectrum of LEs is required, then, for this type of forced oscillator, the second LE can be estimated from the dominant LE and the measured damping value.

These checks serve to illustrate the potentially sensitive nature of some of these computations and the importance of making a variety of diagnostic analyses (especially for noisy, higher-order systems) before firm conclusions can be drawn. However, the largest LE again shows a close correlation between theory and experiment.

A couple of final points here are that a strange attractor may not necessarily be chaotic (Grebogi et al., 1984; Ditto et al., 1990b), and so in this case, the dimension calculation is necessarily augmented by the LE computation. Many of the LE algorithms based on the analysis of experimental data are not appropriate for characterizing periodic behavior since local behavior is not naturally interrogated (this is not a problem for chaos). We also note that if more than one LE is positive (in a high-order system) then the motion is termed hyperchaos (Moon, 1992). Lyapunov exponents will be scrutinized further in Chapter 13.

9.7 Unstable Periodic Orbits

Apart from having a fractal dimension and exhibiting extreme sensitivity to initial conditions, one of the characteristics of a chaotic attractor is the fact that there are a large number of unstable periodic orbits embedded within it. In fact, it can be envisioned that a chaotic trajectory spends most of its time traversing between, and about, the various unstable periodic orbits.

It is possible to characterize certain unstable orbits (and their influence on their surrounding neighborhood) embedded in a chaotic attractor by adapting a least squares fit approach (Lathrop and Kostelich, 1989). This approach, which involves approximating Equation (3.47), will be revisited in a later chapter with regard to stable fixed points. Because points in the Poincaré section are encountered sequentially, they have a specific time label associated with them, and hence, it is relatively simple to identify recurrent points. This is achieved by building up a histogram of highly visited areas. Given a region in state space, it is possible to track those points that iterate back within a small normalized distance ϵ after, say, p cycles. We can then label these as (p, ϵ)-recurrent points. By repeatedly acquiring data based on this type of near-periodic behavior, it is possible to use a linearized least squares approach to construct a local linear approximation to the Poincaré map.

That this kind of nearly periodic behavior occurs in the experimental Duffing system can be observed in a number of ways. First, the chaotic time series (Figure 9.1) clearly shows a strong influence of the forcing period (≈ 1 second in this case) as well as some almost period-five motion between about 30 and 50 seconds in part (a) and between 10 and 20 seconds in part (b). Actually watching this kind of oscillation in the lab, the observer may be temporarily convinced that a stable periodic steady state has been reached, only for a burst of highly random behavior to occur, perhaps followed by another type of recurrent behavior (not necessarily with the same period). It is interesting to note that the broadband power spectrum of this time series (Figure 9.3(a)) includes a spike at the forcing frequency. Whereas the peaks in the corresponding autocorrelation (Figure 9.8) give a stronger indication of some period-five recurrent behavior.

By conducting a study of recurrent behavior, it was found that in the vicinity of $(X, \dot{X}) = (-1.33, 0.55)$, a cluster of $(1, 0.05)$-recurrent points were suggestive of an unstable period-one orbit. To confirm this, all points falling within a normalized distance of $\delta = 0.03$ of this point were stored along with their images and pre-images. The least squares fit was used to construct an approximation to the local Poincaré map. In fact, a more sophisticated analysis, based on Karhunen–Loeve decomposition, indicated that the data sets lay in approximately one-dimensional linear spaces (Bayly et al., 1994). The directions of these manifolds were then

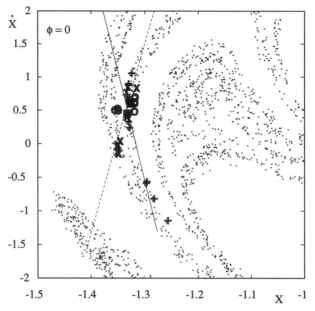

Figure 9.11: A close-up view of part of the experimental chaotic attractor, with the fitted eigenvectors.

used as a basis on which to project the original data, and a simple one-dimensional linear fit was used to estimate the CMs.

Figure 9.11 shows this approach. The recurrent data are based on 14 Poincaré points (○) close to the (1, 0.05)-recurrent point (−1.33, 0.55) and are shown together with their pre-images (×) one cycle earlier and their images (+) one cycle later. They are used to estimate CMs of 2.9 and −0.2 with the corresponding directions (eigenvectors) giving an estimate of the tangent vectors to the stable (dashed) and unstable (lines) manifolds. The route from pre-image to Poincaré point to image can thus be followed with the CM of 2.9 causing a rapid local divergence from the saddle point. Overall the divergence theorem is still satisfied, i.e.,

$$|CM_1||CM_2| = 0.2 \times 2.9 = e^{-4\pi\zeta/0.84}, \tag{9.8}$$

which implies a damping ratio in the range $\zeta \approx 0.036$, a not unreasonable estimate. Clearly, a number of pre-image points very close to the stable manifold allow the nearly periodic behavior to show itself.

The presence of these types of unstable periodic orbits laid the groundwork for the idea of controlling chaos (Ott, Grebogi, and Yorke,

1990). This pioneering approach, which has resulted in considerable research, is based on exploiting some features of chaos to produce a more desirable response. For example, suppose that a periodic response is desired rather than the chaos currently exhibited. Owing to the wandering nature of the behavior, the trajectory passes close to a specific unstable fixed point. Sensing this, an external perturbation could be applied such that the trajectory gets nudged onto the stable manifold (recall Figure 3.4). This would cause the periodic orbit to persist at least until sufficient noise caused the system to be swept back into chaos. However, repeated application of this procedure, which is after all the essence of control theory, will allow the stabilization of an unstable periodic orbit using a very *small* change to the system. Furthermore, in principle, any other unstable orbit could subsequently be stabilized in a similar manner (Ditto, Rauseo, and Spano, 1990a; Dressler and Nitsche, 1992; Bayly and Virgin, 1994). The trajectory can also be targeted to minimize the delay time before the trajectory enters the local vicinity of the fixed point to be stabilized. Thus, for example, extending the periodic regime of a system ordinarily in chaos can be achieved with relatively little effort.

This approach works especially well in numerical simulations or, for example, lasers where there are high signal-to-noise ratios (Roy et al., 1992), but it has less application in mechanical systems where conventional control theory has been applied very successfully. Moreover, a number of practical problems are typically encountered in mechanical systems making it difficult to suddenly change an accessible parameter of the system (Bayly and Virgin, 1993b). Furthermore, the practical motivation, or desirability, for periodic rather than chaotic behavior in mechanical systems is not entirely clear.

Later (in Chapter 14), we shall take a close look at just how the regular motion of the type shown in Figure 8.4 changes into the irregular motion of the type shown in Figure 9.1 (i.e., the transition to chaos).

9.8 Transient Chaos

The randomlike chaotic behavior described in this chapter may also appear for a limited (and perhaps very long) duration prior to settling onto a periodic attractor (Grebogi, Ott, and Yorke, 1983; Tel, 1990; Todd, 1996). This type of transient chaos (obtained under nominally stationary

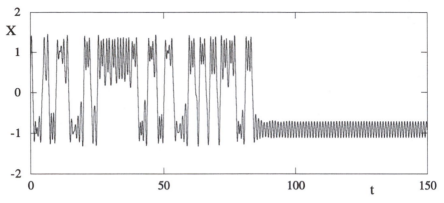

Figure 9.12: An experimental time series showing transient chaos. $F = 0.082$; $\Omega = 0.8$.

conditions) is illustrated in Figure 9.12. The forcing parameters were set at $F = 0.082$, $\Omega = 0.8$ (i.e., a somewhat smaller amplitude than that used to obtain Figure 9.1). The initial cross-well behavior had persisted for about 2 or 3 minutes prior to the start of data acquisition (at $t = 0$), and thus for this (arbitrary) initial condition, the transient motion lasted for about 250 forcing cycles before settling to the small-amplitude periodic attractor in the left-hand well. This is considerably longer than a typical transient, which might last for a dozen cycles or so, depending on various parameters including the damping ratio (see Figure 2.5).

A Poincaré map (not shown) extracted during this transient cross-well behavior reveals a structure not dissimilar to that shown in Figure 9.4(a). A number of other experimental runs were conducted that revealed the arbitrary length of these types of transients, and a number of scaling laws have been developed based on extended numerical simulations (Grebogi, Ott, and Yorke, 1986a). The mechanism underlying this behavior is the appearance of fractal basin boundaries, a subject that will be investigated more fully in Chapter 14 on global issues.

Chapter 10

Escape from a Potential Energy Well

10.1 Introduction

In this chapter we go back and reexamine the transition between the periodic behavior described in Chapter 8 and the thoroughly non-linear chaotic behavior described in Chapter 9. This is an interesting aspect of the double-well oscillator (or indeed any system with a hilltop): Trajectories, initially contained within a single potential energy well, might *escape* over a nearby local maximum (hilltop). In the double-well case, this means that the motion may traverse through to the adjacent well. Clearly, this is a situation completely alien to the confined, parabolic-well, linear oscillator. Dynamical systems characterized by this possibility of escape from a local potential energy well occur in many physical problems including a rigid-arm pendulum passing over its inverted equilibrium position (Baker and Gollub, 1996), snap-through buckling in arch and shell structures (Bazant and Cedolin, 1991), capsizing of ships (Virgin, 1987), and the toppling of rigid blocks (Virgin, Fielder, and Plaut, 1996).

The escape of trajectories from a local minimum of an underlying potential energy function is essentially a transient phenomenon. Given a single-degree-of-freedom system at rest in a position of stable equilibrium, it is often desirable to find the range of harmonic excitation that causes the subsequent motion to overcome an adjacent barrier defined by the limit of the catchment region (basin of attraction) surrounding the minimum. Escape occurs as the motion within the potential well grows "large enough." This is clearly more likely to occur when the forcing magnitude is "large" in relation to some system characteristic. However, even in linear dynamics, the response of a sinusoidally forced system will be magnified close to resonance as we have seen. In nonlinear systems, the size of the basins of attraction surrounding an attractor depend crucially on certain system parameters. Nonstationary changes are incorporated in this chapter to simulate quasi-steady escape.

First, the unforced system is used as an introduction to escape based on initial conditions. Second, a slowly evolving harmonic excitation is applied to the system. The evolution is achieved by changing the forcing amplitude or frequency very slowly, either in an increasing or decreasing manner. In this way transients are minimized such that the evolving trajectory remains "close" to the underlying steady-state solution. Third, the harmonic excitation is suddenly applied (switched on) while the system is at rest, thus generating significant initial transient behavior. In a sense, this can be described as the "worst-case" scenario (akin in some ways to the step response of a spring–mass–damping leading to an overshoot) and is important because a system initially at rest is commonly encountered in many engineering applications (e.g., a structure suddenly subject to an earthquake excitation). Figure 2.5 shows examples for a linear system (initially at rest) that obviously doesn't allow the possibility of escape from its (parabolic) potential well.

10.2 An Asymmetric Potential Energy Shape

In an earlier chapter of this book it was shown that one side of the potential energy shape could be rescaled and conveniently studied in the form

$$\ddot{x} + \beta\dot{x} + x - x^2 = F\sin(\omega t + \phi), \tag{10.1}$$

such that motion may escape to infinity (rather than the adjacent well in Duffing's equation) as x becomes large and positive (see Figure 4.1). There are certain practical advantages for escape to be unambiguously single-sided, and in some sense it is also more generic and typical to have a system that is not perfectly symmetric (e.g., an anharmonic oscillator)

(Huberman and Crutchfield, 1979; Linsay, 1981; Raty, von Boehm, and Isomaki, 1986; Kao, Huang, and Gou, 1988; MacRobie and Thompson, 1991; Rega, Benedettini, and Salvatori, 1991). In this case, escape has profound implications for global stability. Using the same theoretical development as in Chapter 5, we can derive a nondimensional equation of motion that represents the behavior of a small mass oscillating on a *cubic* potential energy shape:

$$(1 + \alpha^2 X^2 (X^2 - 1)^2) X'' + 2\zeta X' + \frac{\mu}{2\alpha} \text{sgn}(X')$$
$$+ \alpha^2 X'^2 X (1 - X)(1 - 2X) + X(1 - X)$$
$$= F\Omega^2 \sin(\Omega\tau + \phi). \tag{10.2}$$

Provided that Coulomb damping is neglected, the forcing is direct (rather than transmissible), and α (where $\alpha = \omega_n^2 x_e / g$) is small, this equation reduces to a form familiarly known as the *escape equation* (Thompson, 1989) (i.e., Equation 10.1). We note at this point that the viscous damping enters the equation of motion in its simplest form (see Section 5.2) and the definition of α is slightly different from the double-well case because the single equilibrium is already at the origin. Physically, the horizontal and vertical distances between the curve's minimum and maximum were 300 mm (the basis of a track "unit" in this chapter) and 54 mm, respectively. However, all numerical simulations to follow (in this chapter) are based on integrating the full equation of motion (10.2), since $\alpha = 0.94$. In any case, there is relatively little computational difference between solving the equation of motion with or without the additional terms.

Given this situation, we now seek to answer the following question: If the initial conditions are zero (the rest state) and the forcing is switched on, what combinations of forcing magnitude (F) and forcing frequency (Ω) cause the system to escape the confines of the local potential energy well? We can also ask how the scenario will change if the forcing parameters are very gradually increased and how quickly will the system escape (measured in real time or forcing cycles).

We thus envision two cases: transient and quasi-steady escape. In keeping with the theme of this book, the description is essentially deterministic. However, escape is clearly related to the problem of first-passage time in stochastics (Arrechi, Badii, and Politi, 1984) in which a particle is subjected to noise until it traverses a certain response level.

10.2.1 The Unforced Case

The nature of the free decay for relatively small amplitudes (using the same cart as for the Duffing system) is used as the basis for matching the viscous damping model between the theory and experiment, and such measurements produced a viscous damping estimate within the range $0.010 < \zeta < 0.014$. For the Coulomb damping coefficient, a flat track section (identical in all respects to the escape potential track) was used at a small inclination to produce a very slow but steady cart velocity, just as before. The latter measurements place the Coulomb damping coefficient in the (very low) range $0.01 < \mu < 0.015$. Typically, damping values near the lower ends of the ranges specified were used in the simulations. Having established estimates for the various system parameters, we can compare the numerical results from the theoretical model to the experimental results. We see a slight change in values from those measured for the double-well Duffing system.

As with the previous system described in Chapter 6, experimentally, a starting gate was used to initiate a series of transients with slightly different initial displacements, all with zero initial velocity. Numerical simulations based on the full model represented by Equation (10.2) were also conducted, and a comparison between theory and experiment is shown in Figure 10.1. Figure 10.1(a) shows two numerically simulated time histories with adjacent initial conditions but spanning the separatrix, such that one trajectory (with $X_0 = -0.53$) fails to overcome the potential barrier at $X = 1$, whereas the other trajectory (with $X_0 = -0.54$) has just enough energy to escape. The experimental results shown in Figure 10.1(b) confirm the location of the separatrix on the position (X) axis and show remarkably similar results. Also, experimentally, $X_0 = -0.53$ results in no escape, and $X_0 = -0.54$ produces escape. The corresponding phase trajectories for the numerical and experimental simulations are shown in Figures 10.1(c) and (d), respectively. The results shown in Figure 10.1 are essentially the same as in Figure 7.7 but with no subsequent return back across the hilltop after escape. Without damping, contours of constant total energy can be used to locate this separatrix as described in Section 7.1.

Some of the minor differences between the numerical and experimental results can be explained by the finite precision of the experimental initial positions with increments of 0.01 nondimensional units.

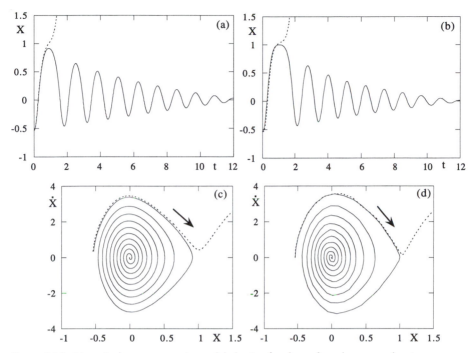

Figure 10.1: Numerical versus experimental behavior for the unforced system, showing bounded and unbounded (escaping) motion. (a) Numerical time series; (b) experimental time series; (c) numerical phase projection; (d) experimental phase portrait. $\dot{X}(0) = 0$ in all cases.

However, effects such as nonlinear damping, overall cart rotational inertia, and deviation from point mass behavior are not considered and may also play a role as discussed earlier. But again, we see a strong qualitative correlation between the two.

10.2.2 The Forced Response

We have already seen that applying a sinusoidal external excitation produces a three-dimensional phase space dominated by periodic attractors. For small forcing amplitudes, the steady-state response of the system is approximately linear, since the system does not extend beyond the locally parabolic shape in the vicinity of the stable equilibrium. For moderate forcing amplitudes, the familiar hysteretic jump in resonance is again encountered and some typical results ($F = 0.055$) are shown in

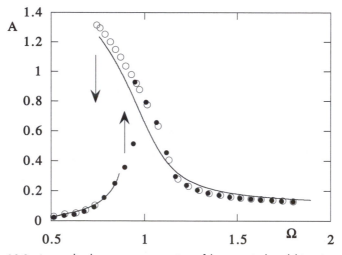

Figure 10.2: An amplitude response comparison of the numerical model (continuous line) and the experiment: ○ – decreasing frequency; • – increasing frequency. $F = 0.055$.

Figure 10.2. Here, the solid line was produced numerically by integrating Equation (10.2) (using a Runge–Kutta algorithm). These responses are steady state, but under the gradual increase in the frequency ratio, the sudden jump (up) initiates a transient that may lead to escape or may converge to the large-amplitude solution. Note that the response tends to zero for low frequencies. This is a consequence of the form of base excitation employed, in contrast to the more typical direct mass excitation of earlier numerical studies. Furthermore, the effect of Coulomb damping overcoming the motion starts to have an influence just below this region. Note the similarity between this response and that shown in Figure 8.5. Other periodic (including subharmonic) responses can be observed for different forcing conditions, for example, a small-amplitude period-two oscillation appears when the system is forced at close to half the natural frequency (see Figure 8.7), although the practical effects of Coulomb damping in the experiment tend to inhibit motion at low frequencies.

The region of hysteresis is again associated with a range of frequencies for which there are multiple solutions, and hence, there is a dependence on initial conditions. Related to this, different steady-state responses may occur for identical initial conditions but slightly different

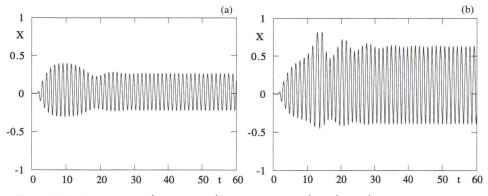

Figure 10.3: An experimental time series showing parameter dependence, that is, a start from rest followed by steady motion for slightly different forcing amplitudes. (a) Small-amplitude motion, $F=0.067$, $\Omega=0.914$; (b) large-amplitude motion, $F=0.069$, $\Omega=0.914$.

parameter values. An example of this is shown in Figure 10.3 where the system is started from rest with slightly different forcing amplitudes (but the same initial forcing phase and frequency). Note that the transient attracted to the resonant branch (Figure 10.3(b)) passes reasonably close to the hilltop at $X=1$ but does not escape. A subsequent small change in a system parameter may result in escape.

10.3 Quasi-Steady Escape

A series of experiments were conducted in which the forcing amplitude was evolved at a user-prescribed rate and at a fixed frequency until escape occurred. Physically, the cart was forced at the desired frequency at low amplitude until a steady, linearlike motion was achieved. The amplitude stepper motor was then computer-controlled to produce the desired amplitude evolution rate. The numerical simulations follow a similar procedure. Figure 10.4 shows the experimental data points resulting from this exercise, as well as a comparison with the numerical model.

At low frequencies, relatively high forcing amplitudes are required for escape, and this escape generally leads to a postjump phenomenon (see Chapter 8). The jump in response can be exhibited under the smooth

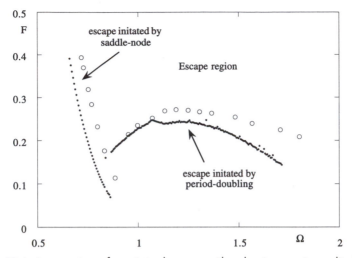

Figure 10.4: A comparison of quasi-steady escape with a slow increase in amplitude while holding frequency constant. •: simulation; o: experiment.

variation of forcing amplitude or (the more familiar) frequency sweep. At higher frequencies (approximately above a nondimensional frequency of 0.9), the peak-to-peak motion gradually grows until escape occurs in which a period-doubling sequence is the underlying instability mechanism. At frequencies near (but still below) resonance, hysteresis effects induce jump-up and subsequent escape. Note, from Figure 10.4, the step change in amplitude (discontinuity) required to produce escape just below the natural frequency. The fold line continues down toward lower forcing amplitudes but the jump does not result in escape (see Figure 10.3). In this sensitive transition region of parameter space (close to resonance), research has shown the possibly indeterminate nature of the postjump behavior (Thompson, 1992); this will be scrutinized in Chapter 14.

Differences between numerically and experimentally generated results can again be attributed to simplifying assumptions made in the theoretical model and nonstationary effects (see Section 8.5). The nondimensional frequency of 1.8 corresponds to the maximum frequency attainable with the experimental forcing apparatus. At high frequencies, the experimental and numerical results have a larger discrepancy. The existence of such bounded and unbounded behavior and their transition can be critical in the stability of real engineering systems subject to

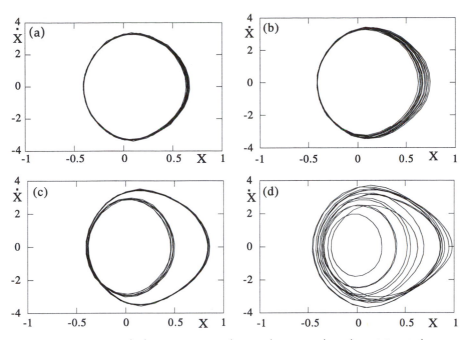

Figure 10.5: Four typical phase projections close to the escape boundary: (a) period-one motion, $F = 0.23$, $\Omega = 1.064$; (b) "sluggish" motion very close to a bifurcation, $F = 0.23$, $\Omega = 1.033$; (c) period-two motion, $F = 0.23$, $\Omega = 1.027$; (d) chaos, $F = 0.21$, $\Omega = 0.94$.

forcing at increasing amplitudes (for example, a vessel at sea subject to increasing wave action (Virgin, 1986)).

Figure 10.5 shows four (steady-state) phase projections as the escape boundary is approached. Subsequent bifurcations occur rapidly, making higher-order subharmonics difficult to locate. However, the period-two oscillation shown in Figure 10.5(c) was stable over a range of (experimentally repeatable) forcing parameters, and the similarity with Figure 8.8(b) is also noted. These specific parameter values can be located in Figure 10.4. The nonlinear asymmetry of the underlying potential energy well is reflected in the (asymmetric) waveform. This subtle scenario in the vicinity of resonance was discussed earlier. Note that the forcing frequency was tuned down slightly in Figure 10.5(d) to observe the chaotic response. This single well chaos is very sensitive and any (small) change or perturbation will typically send the system into cross-well chaos (if an adjacent well exists). We also observe how the dynamics become quite sluggish in the vicinity of a bifurcation (Figure 10.5(b)), which is entirely to be expected and occurs in numerical simulation as well (Thompson, 1989; Seydel, 1991).

10.3.1 The Period-Doubling Cascade

The quasi-static change of a system parameter allows local bifurcations to be realized on the route to escape (if escape occurs at all). The saddle-node associated with the resonant jump phenomenon and the flip bifurcation initiating a period-doubling sequence are the two generic instabilities for this type of system, as shown in Figure 10.4. The jump may occur without escape (this is essentially the situation shown in Figure 10.3(b)). The period-doubling cascade is a particularly intriguing feature in nonlinear dynamics and contains some very interesting scaling behavior (Feigenbaum, 1978). The onset of each subsequent period-doubling bifurcation takes place at a constant geometric rate as a function of the control parameter. If the control parameter is λ, then

$$\frac{\lambda_n - \lambda_{n-1}}{\lambda_{n+1} - \lambda_n} \to \delta, \tag{10.3}$$

as $n \to \infty$, where $\delta = 4.6692016$. What is so remarkable is that this is a *universal* constant for a great variety of sequences found in various maps and flows (May, 1976). It is also a clue to other subtle scaling behavior not uncommon in nonlinear systems. The period-doubling sequence is actually associated with both the saddle-node and flip depending on which way the system is evolving, (i.e., the bifurcation occurs as a simultaneous flip bifurcation for the period-one cycle (CM $= -1$) and a saddle-node bifurcation of the period-two cycle (CM $= +1$)). This *Feigenbaum* scaling (Equation 10.3) makes observing and measuring higher-order subharmonics exceedingly difficult in an experimental context.

Although these bifurcations of periodic behavior comprise the start of the escape scenario, the actual mechanisms of escape are essentially global in nature and will be dealt with in more detail in Chapter 14.

A schematic of bifurcations in the response of this oscillator, based on the work of Thompson (Thompson, 1989; Ueda et al., 1990), is included as Figure 10.6. This three-dimensional diagram shows how many of the oscillations appearing so far, and the general evolution

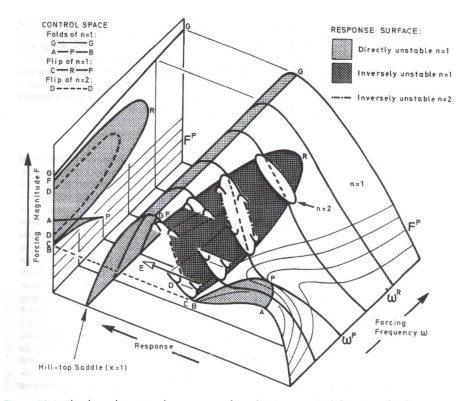

Figure 10.6: The three-dimensional response surface showing generic bifurcations for the escape equation. Reproduced with permission from "Chaotic phenomena triggering the escape from a potential well" (Thompson, 1989).

through increasing forcing, can be envisioned by progressing through the response via horizontal slices. We can also thus consider Figure 8.10 as a simplified (rotated) local view of Figure 10.6 in the vicinity of resonance for moderate levels of forcing.

10.4 Transient Escape

Although there are a great many practical situations in which a control parameter is gradually changed, another important category includes those systems at rest or in a quiescent state that are subject to a suddenly applied load. The limiting case here is an impulse input or shock. But

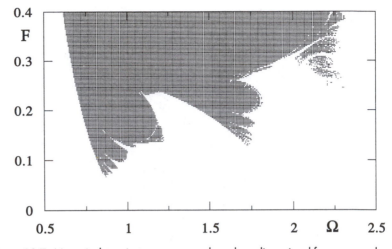

Figure 10.7: Numerical transient escape over a broad nondimensional frequency and amplitude range. Initial conditions are (0, 0).

due partly to practical mechanical limitations, we shall look at a system with zero initial conditions and subject to a sine wave (with a zero phase angle). In this case significant transient motion is induced and escape may well occur long before any kind of steady-state behavior has been approached. Since the origin is usually some distance from a steady-state solution (periodic attractor), it is often the case that this situation can be considered a severe, or even worst-case, escape scenario (other, of course, than the application of a massive step input, which would cause immediate escape under any circumstance).

We first consider this kind of transient escape from a numerical perspective using Equation (10.2). A series of start-ups from rest ($X(0) = \dot{X}(0) = 0$) were simulated at varying forcing amplitudes and frequencies. In Figure 10.7 each simulation that results in escape is represented by a dot. A numerical simulation was started from every point in this rectangle with about 40,000 simulations in all (of which about 22,000 did not lead to escape). Note the complicated nature of the boundary separating escape and no-escape grid points in parameter space and the intricacies of frequency and amplitude combinations that produce narrow regions of bounded motion around the resonant frequency. The lower bound for escape in terms of forcing amplitude bears some resemblence to the curves in Figure 10.4.

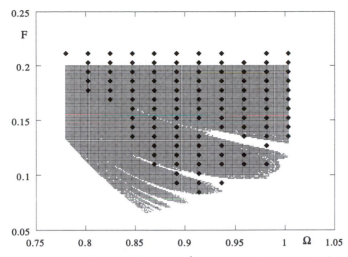

Figure 10.8: Experimental transient data points ◆ superimposed over numerical results in the vicinity of main resonance. Initial conditions are (0, 0).

The experimental analogy of this was achieved, albeit over a necessarily more narrow range of forcing parameters (Figure 10.8). The choice of zero initial conditions is, of course, the easiest to accomplish experimentally. This region was chosen because of the interesting behavior partly triggered by resonant hysteresis effects. Over 200 initial conditions were examined experimentally, and each initial condition was observed at least twice to confirm escape or no-escape behavior. In fact, this last point is a very subtle issue. Although it should be reasonably easy to reproduce the same nominal experimental conditions, the transition between different types of behavior (between escape and no-escape in this case) is very sensitive. Again, at lower frequencies within this range, the experimental system requires larger forcing amplitudes to initiate escape (there is a little start-up transient in the motor itself). Agreement between theory and experiment improves above the resonant frequency. There is good qualitative agreement in general, although there is some difference between the two sets of results at lower frequencies. This discrepancy (which is more apparent using the close-up view of Figure 10.8) is probably caused in part by the difficulty of modeling the actual damping present in the system.

Furthermore, delineating the escape boundary focuses on the most sensitive area. Two representative experimental time series, relating

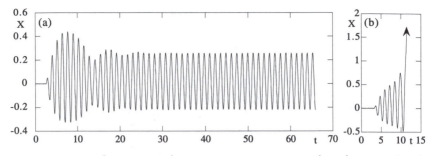

Figure 10.9: Typical escaping and nonescaping trajectories. Initial conditions are (0, 0); $\Omega = 0.88$; (a) $F = 0.102$, (b) $F = 0.104$.

to adjacent parameter values but just spanning the escape boundary, are shown in Figure 10.9. They have the same initial conditions and forcing frequency but a very small difference in their forcing amplitudes. This is very similar to the situation in Figure 10.3 (but with larger forcing amplitudes), and now the alternative outcomes correspond to escape versus no escape as opposed to small versus large response amplitude. Nevertheless, both of these scenarios stand in marked contrast to a typical linear system where small changes in parameters generally lead to small changes in the response with a simple scaling. This uncertainty in outcome will be considered in more detail in Chapter 14.

While the resolution of the experimental initial conditions attempted is necessarily crude, narrow regions of no-escape behavior are observable in the experimental data (Figure 10.8; for example at $\Omega = 0.94$, $F = 0.1$ and at $\Omega = 0.96$, $F = 0.12$). This latter finger of no escape is somewhat counter-intuitive in the following sense. Suppose the forcing frequency is held fixed at $\Omega = 0.94$ and motion is initiated from the rest state with a forcing amplitude $F = 0.07$. The amplitude of the resulting transient is not large enough to traverse the potential energy hilltop. Increasing F slightly (to approximately 0.095) and restarting the experiment (again from rest) results in escape. With $F = 0.1$ there is no escape; with $F = 0.12$ we have escape; with $F = 0.135$ no escape; and with $F = 0.16$ and greater again results in escape.

It can also be noted that each data point corresponding to escape has associated with it an escape time (i.e., a transient length). For example, a point in parameter space deep in the escape region will tend

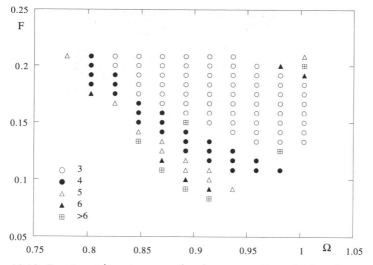

Figure 10.10: Experimental transient escape lengths corresponding to the data of Figure 10.8.

to escape relatively quickly, perhaps during the first forcing cycle. Conversely, data points close to the transition between escape and no escape tend to exhibit relatively long transients. Numerical simulations show the highly complex nature of these escape times and their relationship with transient chaos (i.e., transients of arbitrary length that eventually settle onto a periodic attractor; see Section 9.8) (Thompson, 1989). Indeed, the fractal nature of the escape boundaries suggest, theoretically, that given sufficient precision, it would be possible to locate a pair of forcing parameters that would result in escape after *any* very large number of transient oscillations. In a practical sense, it is assumed that escape will not occur if it hasn't already happened after, say, 50 cycles.

Figure 10.10 shows the experimental escape data but with transient lengths attached. We can thus see how the specific case shown in Figure 10.9(b) would provide a data point corresponding to about 6 forcing cycles and labeled correspondingly. Some representative numerical results are shown in Figure 10.11, where a simplified version of the escape Equation (10.1) incorporating transmissible forcing was numerically integrated for a regular grid of forcing parameters. Hence, this plot is based on a slightly different model to Equation (10.2), which was used for Figure 10.7. In this figure the shading scheme corresponds

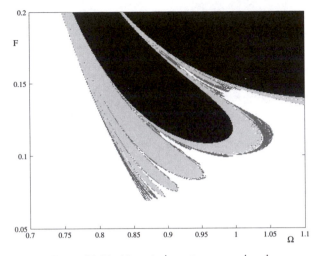

Figure 10.11: Numerical transient escape lengths.

to escape within the following number of forcing cycles:

- black: less than 2,
- light gray: between 3 and 7,
- dark gray: between 8 and 50,
- white: no escape (or more than 50).

The results are qualitatively similar but with generally longer transient lengths, again a discrepancy most likely attributable mainly to damping. The areas of contiguous shade occur in similar shapes with the longer transients tending to be grouped close to the escape boundary. For lower values of damping, the escape area shifts vertically down, a not altogether surprising outcome given the relation between damping and (resonant) amplitude response.

An approximate criterion for escape based on an energy threshold has been suggested (Virgin, 1988; Virgin et al., 1992) that may provide some practical insight to this phenomenon, but clearly the detailed complexity exhibited in the space of forcing parameters (and initial conditions) precludes much hope of obtaining accurate analytic solutions.

An interesting problem, related to this escape scenario, is the rocking block (Virgin et al., 1996). Take a rectangular block that can rotate about either corner (but not slide or bounce) in the plane and subject its foundation to lateral shaking (harmonic). Then determine the range of forcing parameters that lead to rocking and perhaps an overturning instability. Despite the absence of a linear natural frequency and, hence, resonance and the piecewise linear nature of the governing equation of motion, the boundary separating overturning from nonoverturning exhibits similar fractal properties to the above escape problem, together with a variety of periodic and chaotic behavior. A practical motivation might be to predict the overturning of slender statues in a museum suffering an earthquake.

Chapter 11

A Hardening Spring Oscillator

11.1 Introduction

So far, emphasis has been placed on a system that is asymmetric and, as motion grows, exhibits a softening spring effect. Despite the fact that an asymmetry also causes a hardening effect in one direction (unlike the symmetrically soft pendulum), the overall motion is characterized by a natural period that grows with amplitude and a frequency response curve that bends toward lower frequencies. The main new phenomenological ingredient to be added in this chapter is to show that a hardening spring effect is quite similar to the softening spring effect in terms of the amplitude response diagram, but the resonant peak tends to bend toward higher frequencies. It will also be shown that it is possible to extract CMs from experimental data and hence predict the onset of a loss of stability. Specifically, we shall investigate how the decay of small superimposed perturbations gives clues about the approach of a saddle-node bifurcation, which results in the amplitude jump at resonance. This will be achieved using a variation of the basic experimental system. The key diagnostic tool to be used here is an application of the local least squares fit to characterize the periodic orbits (fixed points). This technique was used earlier to study unstable periodic orbits within a chaotic attractor, whereas here it is used for stable fixed points as

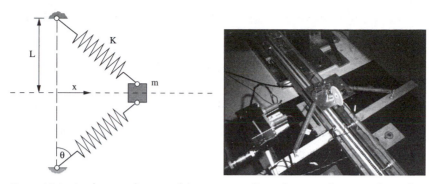

Figure 11.1: A schematic diagram of the geometrically nonlinear oscillator, together with a photograph of the experimental system. The supports of the springs are excited by a harmonic displacement in the *x* direction.

a function of a changing parameter to infer the approaching loss of stability.

The experimental system introduced earlier may be altered to make the potential energy curve steeper than the parabola of a linear system (e.g., $A > 0$, $B = 0$, $C > 0$ in Equation 4.1). Instead, however, we construct a horizontal flat track on which the same cart runs. Now, we attach two linear (helical) springs to the sides of the cart such that the elasticity of the springs rather than gravity provides the restoring force. This is shown schematically in Figure 11.1. To obtain a hardening spring characteristic, the springs are kept in tension, and since the motion takes place in a direction perpendicular to the initial plane containing the springs, it is a geometric effect that causes the stiffness to be a (nonlinear) function of position (see Figure 11.1).

11.2 Mathematical Modeling

Newton's laws applied to the two spring forces acting on the mass show that

$$m\ddot{x} = 2T \cos \theta, \tag{11.1}$$

where T is the tension in the springs and θ is the angle between the springs and the horizontal (Arnold and Case, 1982; Forbes, 1989).

153

Alternatively, use can be made of Lagrange's equation to obtain the governing equation of motion of the (now damped, driven) mass:

$$\ddot{y} + \beta \dot{y} + \left[1 - \frac{\lambda}{\sqrt{1 + y^2}} \right] y = \frac{F}{L} \Omega^2 \cos(\Omega \tau), \qquad (11.2)$$

where y is the position nondimensionalized relative to the initial (stretched) length of each spring ($y = x/L$), F is the forcing magnitude, Ω is the ratio of the forcing frequency ω to the natural frequency (measured as $\omega_n = 1.1$ Hz), and $\tau = \omega_n t$ (also the basis for the time derivatives). A further parameter is introduced, λ, which relates the unstretched length of the spring to its length when $y = 0$ (i.e., $\lambda = l_r/L$).

This configuration proves to be quite suitable for obtaining a restoring force with zero linear stiffness. This is analogous to the axially loaded strut with an end load exactly equal to its elastic critical load (as discussed in Chapter 4) and has also been studied extensively as a model of certain electric circuits (Hayashi, 1964; Ueda, 1980).

11.3 Amplitude Frequency Response

We encountered the possibly multivalued nature of nonlinear resonance in Chapter 8 and will demonstrate a similar (but interestingly different) behavior here. Assuming purely linear viscous damping and using the log dec approach, a damping value of $\beta = 0.06$ was estimated from a small-amplitude experimental free decay. Furthermore, with a forcing amplitude F of about 13 mm and length L equal to 219 mm, a nondimensional forcing magnitude F/L of 0.06 was selected (Murphy et al., 1994), and this value will be used throughout this chapter. We can conduct a standard exercise for this oscillator by sweeping through forcing frequency and plotting the resulting resonance response diagram as shown in Figure 11.2

For the value of the forcing amplitude used here, forcing frequencies in the vicinity of $\Omega \approx 1.1$ ($\omega = 1.21$ Hz, $\tau = 0.83$ s) result in coexisting periodic attractors and hence a dependence on initial conditions. This can be directly compared with Figures 8.5 and 10.2. For example, Figure 11.3 shows two coexisting attractors when $\Omega = 1.1$, which can

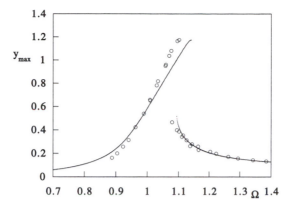

Figure 11.2: Amplitude response diagram showing the hardening spring effect and hysteresis. The open circles are experimental measurements and the lines are from numerical simulation. $F/L = 0.06$.

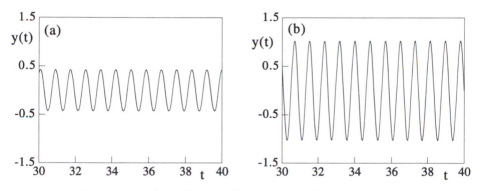

Figure 11.3: Coexisting periodic oscillations in the main resonant hysteresis region (experimental data). $\Omega = 1.1$.

be termed nonresonant for the small-amplitude motion and resonant for the large-amplitude motion, analogous to Figure 10.3. Which of these steady-state motions persists after the decay of transients depends on the initial conditions: At this value of the parameters the two attractors are roughly equally dominant. To determine these relationships more fully, a global analysis must be conducted to determine the basins of attraction. This will be deferred to Chapter 14.

However, as we have noted, it is also quite likely that in a realistic situation one of the parameters might be slowly changed. This would then result in a jump up (or down, depending on the direction of sweep) at a saddle-node (cyclic fold bifurcation). It is the *approach* to this sudden qualitative change in behavior that will be subject to further scrutiny.

11.4 Stability Analysis

In order to investigate the stability of the periodic attractors, we will induce small perturbations as the frequency of the forcing is gradually reduced toward resonance and the jump to the larger amplitude solution is approached (which occurs at approximately $\Omega = 1.05$).

It is again convenient to reduce this three-dimensional flow to a two-dimensional discrete map by using Poincaré sampling. Hence, examining the stability of the periodic solution is equivalent to considering the stability of a fixed point in a plane as described in Chapter 3. After the application of a perturbation, the ensuing transient will either decay back to the original attractor or go elsewhere. This latter behavior will occur if the fixed point has lost its stability or the perturbation was sufficiently large that a remote, but coexisting, solution was approached. The choice of the Poincaré sampling section in phase space is again determined by the forcing phase.

Since the applied perturbations are relatively small, we can assume that the transient trajectory thus created is governed by a local linear approximation. That is, the full map

$$x_{n+1} = g(x_n) \tag{11.3}$$

is replaced by

$$x_{n+1} \approx [A](x_n) + b, \tag{11.4}$$

where the $N \times N$ matrix $[A]$ is the local Jacobian matrix of the Poincaré map ($N = 2$ in our case). The stability of the map is determined by the nature of the eigenvalues. This is quite analogous to the fundamental notion of stability of equilibrium. However, we now have CMs that must remain within the unit circle for stability (see Figure 3.6).

11.4.1 Experimental and Numerical Approach

The determination of CMs from experimental or numerical data is quite straightforward. A perturbation is applied to a system undergoing steady-state oscillatory motion. Following the perturbation, M Poincaré points, $\mathbf{x}_n$, in phase space and their images, $\mathbf{x}_{n+1}$ (samples taken one cycle later), are recorded. This process is repeated and the samples are placed into matrices $[W]$ and $[V]$. Incorporating a constant matrix $[B]$ results in

$$[V] = [A][W] + [B], \tag{11.5}$$

which can be solved for matrices $[A]$ and $[B]$ provided enough samples are taken. In general, many perturbations are applied, and many Poincaré points are acquired such that Equation (11.5) is overconstrained but can be solved in a least squares sense. Using more data points in the experiment tends to reduce the effect of noise and increase robustness. Physically, then, the Jacobian $[A]$ contains information regarding the behavior of local transients, and $[B]$ contains information regarding the location of the fixed point. Therefore, it is the eigenvalues of $[A]$ (the CMs) that are of most interest from a stability point of view.

Utilizing this technique for the hardening spring oscillator involves constructing a small braking mechanism capable of applying small disturbances to the periodic steady-state motion. The brake was controlled by a solenoid, which was activated when a marker on the forcing mechanism interrupted a photodetector. The length of contact between the cart and the brake was used to vary the magnitude of the induced perturbation. Poincaré points were acquired, and time-lag embedding was used to extract the discrete state of the system. This can be considered as a local analogy to the global method of stochastic interrogation to be considered later.

Figure 11.4 shows a typical (nonresonant) periodic oscillation being subjected to a perturbation (near $t = 33$) with the resulting induced transient. This plot is based on numerical data and relates to a frequency ratio just above resonance. Figure 11.5(a) shows one data point extracted from each forcing cycle as a function of time, and Figure 11.5(b) shows the corresponding two-dimensional mapping. Since the time step is taken as an exact fraction of the forcing period (e.g., $\Delta t = T/50$) each data point and its image one cycle later ($\pi/25\Omega$, e.g., 50 time steps) provides a Poincaré mapping. Thus we can observe the decay of the transients and characterize this decay (using the least square fit described earlier) to extract CMs. Note the analogy with a flow in the

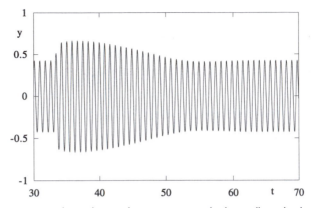

Figure 11.4: Numerical simulation showing a perturbed, small-amplitude time series $(\Omega = 1.1)$.

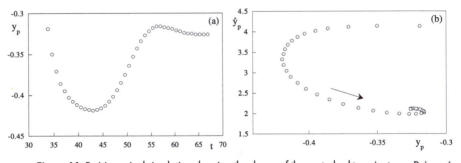

Figure 11.5: Numerical simulation showing the decay of the perturbed transient as a Poincaré map.

vicinity of equilibrium, shown for a linear system in Figure 2.5. When the transient has decayed, another perturbation is applied and the resulting ensemble treated in the aforementioned least squares sense. The control parameter (forcing frequency) is incremented, and the procedure is repeated.

An equivalent example based on experimental data is shown in Figure 11.6(a). Also shown in Figure 11.6(b) is what happens if the applied perturbation is sufficiently large for the trajectories to exit the local basin of attraction and move to the large (resonant) solution (these data were actually acquired for a slightly different set of the parameters but the trend is clear). In general the perturbation (induced by the brake mechanism) is applied at an arbitrary forcing phase and for an arbitrary

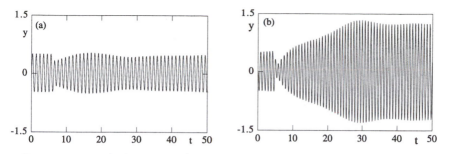

Figure 11.6: Perturbed experimental time series ($\Omega = 1.1$). Part (b) shows a relatively large perturbation that causes the coexisting attractor to be revealed.

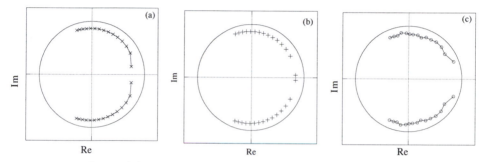

Figure 11.7: Evolution of the eigenvalues (CMs) in the complex plane as a function of forcing frequency with Ω starting from 1.45 (left-most eigenvalues). (a) Numerical: $\times$, (b) analytical: $+$, over the same range of Ω, (c) the corresponding experimental data: $\circ$.

duration. Hence, when incorporated into the least squares scheme the local (linear) transient behavior can be thoroughly investigated.

The procedure was repeated at this and other frequencies to produce an effective root locus of periodic stability as a parameter was changed. These results are shown in Figure 11.7 as the forcing frequency is reduced toward the saddle-node bifurcation. For completeness, some analytic stability results (based on conventional Floquet theory) are included as a comparison (Murphy et al., 1994). The progress toward coalescence near $+1$ on the real axis is clearly seen.

The left-most eigenvalues in each plot correspond to a frequency of $\Omega = 1.45$ with each subsequent calculation at an increment of $\Delta\Omega = -0.0254$. The coalescence on the real axis occurs immediately prior to the instability. Numerical difficulties prevented results beyond $\Omega = 1.094$, but results were obtained quite close to the jump as indicated in Figure 11.2. We can also make an independent check on the magnitude of the CMs using the constraints placed on the system by the divergence theorem. Based on $\beta = 0.06$, we thus anticipate (and obtain) the CM magnitude of $[\exp(-2\pi(0.06)/1.45)]^{1/2} = 0.88$ at the start of the root locus, together with a mild reduction as the frequency is further reduced.

11.4.2 Analytic Approach

An analytic solution to the governing equation of motion can also be attained using a number of different techniques. As a typical example, we will use the method of harmonic balance augmented by some Floquet theory to obtain some approximate solutions to the hardening spring oscillator.

To facilitate an analytic solution, the nonlinear stiffness in Equation (11.2) is expanded as a polynomial (Steidel, 1989) valid over the range $|y| \leq 1.5$ to give an approximate equation of motion:

$$\ddot{y} + \beta \dot{y} + (Ry + Sy^3 + Ty^5) = \frac{F}{L}\Omega^2 \cos(\Omega\tau). \tag{11.6}$$

The experimental system gave values of $\beta = 0.06$, $R = 1.0$, $S = 0.3689$, $T = 0.071$, and $F/L = 0.06$.

A periodic solution to Equation (11.6) is assumed to take the form

$$y_s(t) = A \cos(\Omega t - \phi), \tag{11.7}$$

which is then subject to a small perturbation:

$$y(t) = y_s(t) + \xi(t). \tag{11.8}$$

This leads to a variational equation that governs the behavior local to the periodic solution and, hence, governs its stability:

$$\ddot{\xi} + \beta \dot{\xi} + \xi(R + 3Sy_s^2 + 5Ty_s^4) = 0. \tag{11.9}$$

The steady-state solution, y_s, is then substituted and after various manipulations and transformations we again end up with a form of Mathieu's equation (McLachlan, 1964):

$$\ddot{\eta} + [\alpha - 2q \cos(2z)]\eta = 0, \tag{11.10}$$

where $\eta(t) = e^{\beta t/2}\xi(t)$ and $z = \Omega t - \phi$. The stability of the solutions of Equation (11.10) depends on α and q, which depend on the original parameters of the problem. Despite its rather benign (linear) appearance, Equation (11.10) is by no means simple to solve (Nayfeh and Mook, 1978). However, it is possible to extract the appropriate CMs, and these are also shown in Figure 11.7(b).

The approach of the instability can be predicted by plotting the phase of the eigenvalue as a function of the control parameter as discussed in Ref. (Murphy et al., 1994). The extraction of CMs and their monitoring in an experimental context prove particularly useful as the technique does not require much a priori information about the system.

11.5 Zero Linear Stiffness – Period-Doubling Revisited

Unlike the gravity-driven cart–track system, the physical system considered in this section has considerable flexibility in varying the form of the restoring force. An example of special interest occurs when $\lambda = 1$. In this case the spring is initially free of axial force, thus rendering the linear stiffness (in the lateral direction) zero. This is somewhat related to an axially loaded column when the end load is the Euler critical load and, hence, the potential energy is flat in the vicinity of the origin (as discussed in Chapter 4).

Consider, then, the purely cubic nonlinear oscillator for which the stiffness is an appropriately truncated expansion of the stiffness from Equation (11.2), and B replaces the forcing magnitude (for consistency with previous work):

$$\ddot{x} + 0.1\dot{x} + x^3 = B\cos t. \tag{11.11}$$

A small amount of linear viscous damping has been added. Fixing the forcing frequency at unity, we can now vary the forcing magnitude, B, as the control.

This oscillator has been the subject of a classic analog computer study and harmonic balance analysis (Hayashi, 1964), and its bifurcational structure has also been intensively studied by Ueda (Ueda, 1980). The incredibly complicated nature of possible responses is apparent from Figure 11.8, reproduced from the original paper. A gallery of

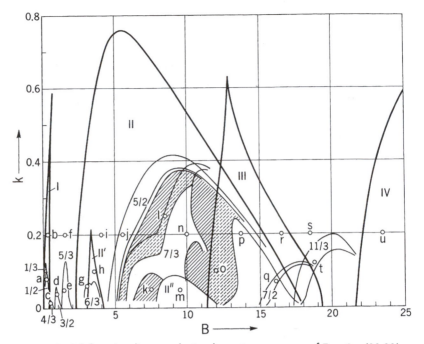

Figure 11.8: A bifurcation diagram plotting the various responses of Equation (11.11) as a function of damping and forcing amplitude. Note: k is used as the symbol for damping in this figure. Reproduced with permission from "Steady motions exhibited by Duffing's equation: A picture book of regular and chaotic motions" (Ueda, 1980).

corresponding phase projections were also illustrated in this reference, and although they are not repeated here, we simply note that the Roman numerals refer to periodic motions with period 2π; the regions indicated by fractions refer to various subharmonic and ultrasubharmonics; and the shaded regions refer to those parameter regions where chaos is present. These regions can then be further refined to distinguish between those regions where a particular response is unique and those where coexisting solutions are present; for example a region with continuous line shading means that chaos is the unique outcome (from any set of initial conditions), unlike the regions shaded by dashed lines in which periodic behavior coexists with the chaos. Higher-order ultrasubharmonic motions are not even included in this figure, and the responses for very light damping become increasingly difficult to identify. Thus, the qualitative nature of the response changes as one of the controls is slowly varied and bifurcations are encountered. The reader might like to contemplate this figure with regard to that obtained using a similar approach on a linear system.

This, and other studies (Parlitz and Lauterborn, 1985), indicate that a period-doubling sequence (for $\beta(\equiv k) = 0.1$) occurs within the range $4 < B < 6$, as shown in Figure 11.9. At this point, it is also worth reiterating the generic nature of this kind of response. For example, almost exactly the same period-doubling tree is displayed by the one-dimensional (quadratic) map

$$x_{i+1} = C - x_i^2, \tag{11.12}$$

including the universal scaling first observed by Feigenbaum (Feigenbaum, 1978).

The changing qualitative nature of the response can now be charted using the behavior of small perturbations from the previous section. The local linear approximation to the behavior can be fit using the least squares approach, and CMs can be extracted from the Jacobian (Virgin and Murphy, 1994). In polar form these eigenvalues can be written

$$\lambda = r e^{\pm i\phi}, \tag{11.13}$$

where $r < 1$ for stability and ϕ is a measure of the frequency of rotation of the transient motion about the fixed point. Two typical perturbations are shown in Figure 11.9 for a period-two subharmonic (every second Poincaré point is extracted in the computation). At $B = 5.0647$, we find $r = 0.532$ and $\phi = 1.04 \approx 2\pi/6$ radians, and, thus, we have roughly six-cycle transients. Again the magnitude of the CM suggests $\beta \approx 0.1$. At $B = 5.3$, we find $r = 0.5318$ and $\phi = 2.098 \approx 2\pi/3$ radians with roughly three-cycle transients. Therefore, although the stability of the response has effectively remained unchanged over this relatively small parameter change, the degree of spiraling has changed appreciably.

This process can now be repeated over a continuous range of forcing amplitudes, B, and the resulting trace can be plotted in the complex plane. Because the flow contracts exponentially (see Section 3.1), the product of the two CMs is a constant for any period-n orbit:

$$\lambda_1 \lambda_2 = e^{-n\beta T}, \tag{11.14}$$

where T is the period of the orbit and β is a damping coefficient. This

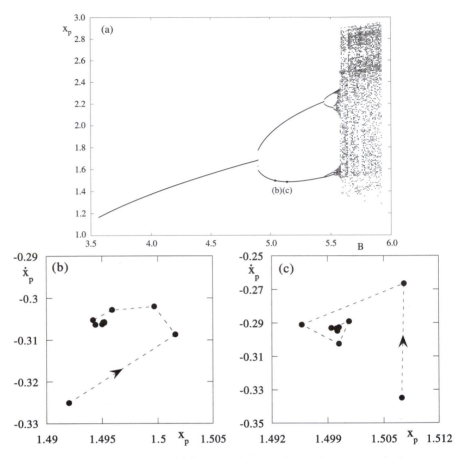

Figure 11.9: (a) A numerical bifurcation diagram plotting the Poincaré displacement x_p versus the forcing amplitude B, highlighting a period-doubling sequence exhibited by Equation (11.11). Also shown are typical transient decay trajectories onto a fixed point: (b) $B = 5.0647$; (c) $B = 5.3$.

product is also equal to the square of the radius

$$\lambda_1 \lambda_2 = r_n^2, \tag{11.15}$$

and thus combining Equations (11.14) and (11.15) gives

$$r_n = \sqrt{e^{-n\beta T}}. \tag{11.16}$$

For the specific parameter values considered here, we find

$$r_n = e^{-n\pi/10}, \tag{11.17}$$

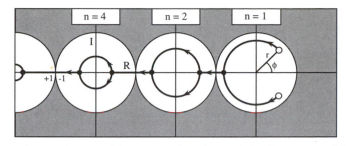

Figure 11.10: A schematic diagram of the transition through period-doubling in terms of the CMs.

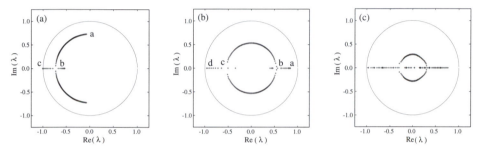

Figure 11.11: The movement of CMs in the complex plane: (a) $n=1$ solution, $B=3.75$ to 4.899; (b) $n=2$ solution over its complete range of existence; and (c) $n=4$ solution, $B=5.45$ to 5.526.

which gives the square root scaling law

$$r_n^{1/n} = e^{-\pi/10} = \text{constant}. \tag{11.18}$$

This behavior is shown schematically in Figure 11.10 for a period-doubling sequence over the first few bifurcations. We note that a reverse sequence is encountered if the control parameter is reduced: The successive bifurcations will then be saddle nodes as the CMs for each subharmonic, transition out of their unit circles via $+1$ on the real axis.

Using this numerical least squares approximation to extract the eigenvalues from the Jacobian leads to the results shown in Figure 11.11.

These figures are clearly analogous to the period-one results shown in Figure 11.7. The initial computation (point a) is for $B=3.575$ for the $n=1$ solution. As the forcing amplitude is lowered (with the forcing frequency held fixed), the CMs rotate on a circle until they coalesce on the real axis at approximately $B=4.77$ (point b). This is quickly followed by one of the CMs exiting the unit circle at -1 ($\phi=\pi$, point c) as

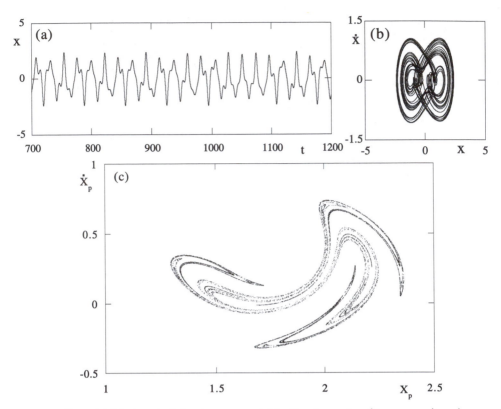

Figure 11.12: A typical chaotic response. (a) Position time series; (b) corresponding phase projection; (c) Poincaré section. $B = 10$.

the period-two solution comes into existence. This behavior is repeated for subsequent flip bifurcations, and the numerical results confirm the scaling law developed earlier (considering for example the radius at $\phi = \pi/2$).

This general approach therefore works well in extracting useful stability information from perturbation-induced transients. This is especially true for experimental data where the least squares approach can be used to reduce the effect of noise. There are, however, some practical issues concerning the optimal size and scheduling of perturbations (e.g., see Figure 11.6).

Before leaving this chapter, we take a brief look at a chaotic response occurring in this system. Figure 11.12 shows a typical times series (a), phase projection (b), and Poincaré section (c) (i.e., the time series sampled at intervals of the forcing period 2π), all computed from a numerical simulation. Note the similarities and differences with the two-well Duffing chaotic attractor shown in Figure 9.4(b). It is also worth noting that this response occurs at a relatively large forcing at which the induced lateral displacements are roughly an order of magnitude greater than the initial length of the spring and, owing to practical limitations, are not the subject of experimental confirmation.

Chapter 12

The Effect of a Stiffness Discontinuity

12.1 Introduction

The systems described thus far have smooth stiffness characteristics where the nonlinearity gradually enters into the picture. However, there is a large class of mechanical problems for which there is some kind of discrete change in the stiffness characteristic. A system may behave in a linear way within a certain subset of the phase space, but overall, the principle of superposition does *not* hold. Examples include various loose structures and components (Peterka and Vacik, 1992; Conner et al., 1997), rotordynamics with rubbing (Ehrich, 1991), backlash in gear mechanisms (Karagiannis and Pfeiffer, 1991), impact oscillators (Shaw and Holmes, 1983; Thompson and Ghaffari, 1983; Bayly and Virgin, 1993b), and rocking blocks (Virgin, Fielder, and Plaut, 1996). The physical paradigm used in the earlier chapters can be amended relatively easily to mimic a typical vibro-impact system and the resulting thoroughly nonlinear behavior typical of such systems. Dry friction, which was encountered earlier, also exemplifies a piecewise linear characteristic (Begley and Virgin, 1997).

Figure 12.1 shows a schematic of the track–cart system with a parabolic track. We have seen in the mathematical modeling chapter how this gives a reasonable approximation to a Hookean spring. A movable barrier with a stiff spring is now clamped at various locations along

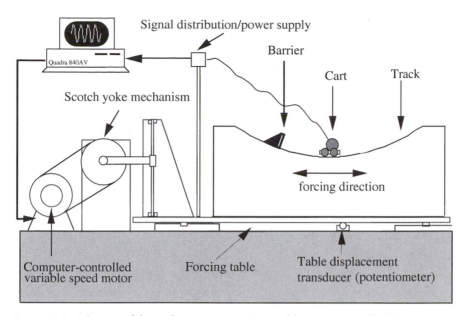

Figure 12.1: Schematic of the track–cart system with a rigid barrier against which the cart impacts (see also Chapter 6).

the track, such that the cart experiences a rapid velocity reversal at contact.

Suppose the cart moves on the track to the right of the barrier. Then the two extremes are: (i) The barrier is placed sufficiently far left of the minimum that the cart does not come into contact with it and (ii) the barrier is sufficiently far to the right of the minimum that the cart remains in contact with the barrier at all times, or at least much of the time. Clearly the most interesting cases are the intermediate ones, and in this chapter we shall briefly consider some free and forced vibration characteristics in terms of the effect of barrier location, again using the forcing conditions as the primary controls.

12.2 Free Response

It is useful (in the usual way) to first consider the undamped motion of the unforced system in terms of the ellipses carved out in the phase plane. The placement of a barrier will effectively clip the phase trajectory

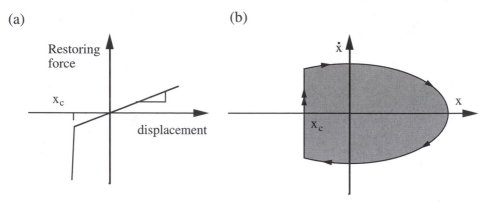

(a)

Restoring
force

x_c

displacement

(b)

$\dot{x}$

x

x_c

Figure 12.2: A schematic of (a) the restoring force, and (b) the resulting truncated ellipse in the phase plane, for an impacting system.

ellipse at the appropriate position. A rigid impact with a purely elastic rebound results in a simple change in the sign of the instantaneous velocity (i.e., the kinetic energy is conserved). Although there is some spring compression during impact (an effect that can be modeled by assuming a bilinear spring characteristic with a discrete change in the stiffness when contact is made), it is assumed that the time of contact is instantaneous, at least in relation to the time spent rolling on the track. This type of stiffness function is shown in Figure 12.2(a), where the very high stiffness of the rebounding spring is added to the gravitational stiffness (acting via the track slope) when $x < x_c$.

Again, we make the assumption that the parabolic track gives a reasonably good approximation to a linear oscillator. This assumption was supported by the fact that the nonlinear terms (see Chapter 5) were negligbly small based on a track shape $(1/2)ax^2$ with $a^2 = 8.25 \times 10^{-5}$ cm^{-2}. In the absence of damping, the phase trajectories without the impact barrier are given by

$$\left(\frac{\dot{x}}{\omega_n}\right)^2 + x^2 = x_0^2, \tag{12.1}$$

where ω_n is the linear natural frequency $(= \sqrt{ga})$ and x_0 is the initial position (release point of the cart) assuming a start from rest (i.e., zero initial velocity). These phase trajectories represent harmonic motion from which the natural period, T_n, can be extracted by evaluating the

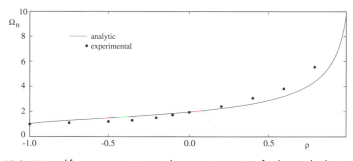

Figure 12.3: Natural frequency variation with constraint position for the purely elastic rebound and the experiment.

closed orbit in the phase plane to give (Jordan and Smith, 1977)

$$T_n = \frac{4}{\omega_n} \int_0^{x_0} \frac{1}{\sqrt{x_0^2 - x^2}} \, dx, \tag{12.2}$$

which gives the expected result $T_n = 2\pi/\omega_n$.

When the impact barrier is placed at location x_c (see Figure 12.2(b)), the period of the motion is thus given by evaluating Equation (12.2) with x_c entering as a limit of integration:

$$T_{n,i} = \frac{2}{\omega_n} \left[\frac{\pi}{2} - \sin^{-1} \left(\frac{x_c}{x_0} \right) \right]. \tag{12.3}$$

For the centrally placed barrier (i.e., in the bottom of the well) we get a natural period of one-half the linear case. In terms of nondimensional frequency this gives

$$\Omega_n = \frac{\pi}{\pi/2 - \sin^{-1} \rho}, \tag{12.4}$$

where $\Omega_n = \omega_{n,i}/\omega_n$ (and where $\omega_{n,i}$ is obtained from the reciprocal of Equation 12.3) and $\rho = x_c/x_0$. Figure 12.3 shows this relation between impact location and the natural frequency. Also shown in this figure are some experimental data points, obviously incorporating some damping effects but agreeing well with the expected analytic trend. More experimental details will be given later with respect to the forced case. Just as with the Duffing track, a more accurate analysis will provide a non-linear equation of motion that admits solutions that are not quite elliptical and will account for a minor adjustment of the experimental–theoretical

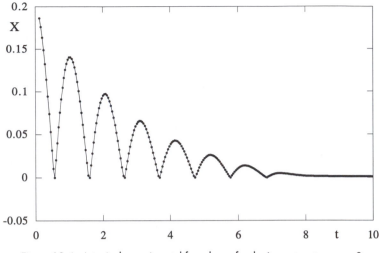

Figure 12.4: A typical experimental free decay for the impact system. $\rho = 0$.

correlation. Another factor in the slight (asymmetric) discrepancy between simulation and experimental results is due to the fact that a small amount of penetration occurs as the spring deflects. For an accurate analysis the system can be studied as a bilinear stiffness, but the assumed pure elastic rebound is good enough here since the compression of the spring is very small compared with the overall motion of the cart.

A typical free decay taken directly from the experimental setup is shown in Figure 12.4 for the case of the impact barrier placed directly at the bottom of the parabola at the position of static equilibrium ($x_c = \rho = 0$). This response is essentially a rectified, exponentially decaying sine wave. Energy is lost both by continuous processes (i.e., during the free motion) as well as by a small amount of discrete damping at impact, which can be modeled in a variety of ways but most popularly with a coefficient of restitution (Todd, 1996).

12.3 Forced Response

We now move on to consider the forced response when the system is subject to horizontal base movement in the usual manner. Initially, attention will be focused on the case $x_c = 0$.

Using the same methods as previously described, the governing equation of motion can be derived as (Todd and Virgin, 1997a)

$$(1+\alpha^2 X^2)\left(X'' + \frac{2\zeta}{\Omega}X'\right) + \frac{\mu}{\alpha\Omega^2}\frac{X'}{|X'|} + \alpha^2 X(X')^2 + \frac{1}{\Omega^2}X = F\sin\tau,$$

(12.5)

where the following nondimensional terms have been introduced: $X = x/x_g$, $\Omega = \omega/\omega_n$, $\omega_n^2 = ga$, $\alpha^2 = x_g^2\omega_n^4/g^2$, $F = f/x_g$, and $\tau = \omega t$. The track term x_g is 0.269 m, a somewhat arbitrary number but taken to be in the vicinity of the double-well track parameter.

For the experimental system, the nondimensional parameter is $\alpha^2 = 0.061$, which suggests the relatively mild effect of any nonlinear deviation from a linear oscillator (but with Coulomb damping) given by

$$X'' + \frac{2\zeta}{\Omega}X' + \frac{\mu}{\alpha\Omega^2}\frac{X'}{|X'|} + \frac{1}{\Omega^2}X = F\sin\tau.$$

(12.6)

To confirm this, a spot check can be made between the amplitude response obtained from the numerical simulation of the full equation of motion (12.5) and the linearized version (12.6). They are shown as bifurcation diagrams in Figure 12.5(a) and (b), together with the corresponding experimental result in Figure 12.5(c). Here, the nonimpacting, measured natural frequency ($\omega_n = 3.14$ rad/s) is used for nondimensional purposes, and thus the primary resonance occurs near $\Omega = 2.0$. The damping values used in the numerical simulation were again extracted by conducting a broad fit from free and forced vibration data: $\zeta \approx 0.1$ and $\mu \approx 0.003$.

The bifurcation diagram confirms the close similarity between the behavior of Equations (12.5) and (12.6), even capturing an interesting appearance of a mild period-two solution for a frequency ratio slightly greater than unity. Coulomb damping likely causes this behavior, although this subtle phenomenon was not observed in the experiment. Another improvement in modeling could perhaps be achieved by adding a small amount of discrete energy loss at contact. This can be conveniently incorporated into the theoretical model using a coefficient of restitution relating pre- and post-impact velocities.

One of the salient features of these bifurcation diagrams is that the primary and secondary resonant peaks are separated by a chaotic

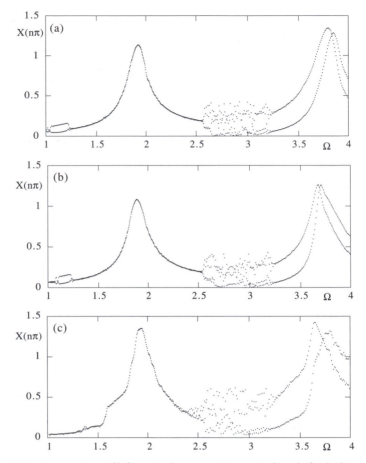

Figure 12.5: A comparison of bifurcation diagrams: (a) numerical results for the linear equation, (b) the full equation of motion, (c) experimental results. $F = 0.07$; $X_c = 0$.

window. In fact, chaotic windows continue to appear between higher resonant peaks and some interesting scaling occurs. In the current treatment, experimental limitations associated with the shake table would not allow investigation very far into this higher frequency regime. Another interesting type of generic behavior found here is the changing nature of the resonant peaks. For example, for frequency ratios up to about 2.5, the behavior is not dissimilar to that of a simple rectified sine wave. However, for higher frequencies (i.e., beyond the first resonant peak) we observe a splitting in the bifurcation diagram. The y axis in these figures represents a Poincaré position at which the surface of

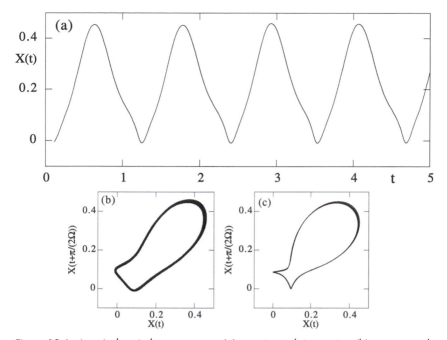

Figure 12.6: A typical period-two response: (a) experimental time series, (b) experimental phase projection, and (c) simulation. $F = 0.07$; $\Omega = 3.5$; $X_c = 0.0$.

section is defined to stroboscopically sample the trajectory. This period-two solution subsumes an impact every *other* forcing cycle. This is in contrast to a period-two motion with alternating high and low amplitudes. Thus a typical time series (for $\Omega = 3.5$) as shown in Figure 12.6 is quite like a triangular wave form, with the cart experiencing fairly uniform velocity magnitude as it travels near the contact. A phase projection using time-delay coordinates is shown also, together with a comparison with the results from a numerical simulation.

Some other responses are shown in Figure 12.7, and they can also be identified from Figure 12.5. The Poincaré section taken from close to the primary resonance gives the single repeating point as expected (Figure 12.7(a)). The corresponding power spectrum displays strong frequency contributions at multiples of the fundamental forcing frequency (the power scale is logarithmic). This is not surprising when one considers the fact that the time series resembles a rectified sine wave. A period-three response is shown in Figure 12.7(c) and (d). Here the underlying waveform experiences two impacts and repeats itself after three forcing

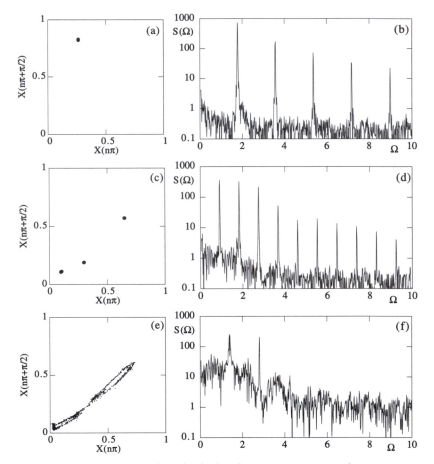

Figure 12.7: Some experimental results displayed as Poincaré maps and power spectra. $F = 0.07$; $X_c = 0.0$; (a) and (b) $\Omega = 1.8$, (c) and (d) $\Omega = 2.78$, (e) and (f) $\Omega = 2.84$.

cycles. This response appears over a narrow window within the chaotic regime. The power spectrum exhibits frequency spikes at one-third and two-thirds of the fundamental in addition to their harmonics.

A typical chaotic attractor is shown in Figure 12.7(e) and (f). The broadband nature of the power spectrum is evident. The single branch, or *finger*, of the chaotic attractor is quite typical for these types of impact oscillator. After each resonant peak (at frequencies of $\Omega = 2n$ in this scaling), the band of chaos is characterized by n fingers. This is shown in Figure 12.8 with two Poincaré sections taken from the experiment with $X_c = -0.1$. Figure 12.8(a) shows a typical chaotic attractor at

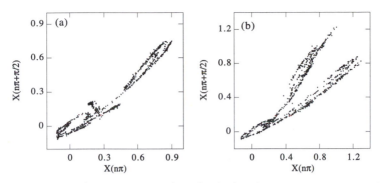

Figure 12.8: Some chaotic experimental results displayed as Poincaré maps. $F = 0.11$; $X_c = -0.1$; (a) $\Omega = 2.35$, (b) $\Omega = 3.8$.

$\Omega = 2.35$, that is, a frequency higher than the main resonant peak, which for this position of the impact barrier occurs at $\Omega = 1.7$. Figure 12.8(b) shows a typical chaotic attractor beyond the next resonant peak. The addition of the second finger is clear, and this sequence continues for higher frequencies. Figure 12.9 shows the time series for these cases (b) and (d), as well as some interchaos periodic behavior. A Poincaré section indicated that part (a) is a period-one oscillation and part (c) is a period-four oscillation. A variety of other periodic and nonperiodic responses were observed.

A great deal of dynamic information is contained in the sequence of impacts. Clearly, the interimpact interval (Slade, Virgin, and Bayly, 1997), for the periodic behavior shown in Figure 12.9(a), is a constant. But the chaotic behavior shown in Figure 12.9(d) can also be reconstructed (see Section 6.5) on the basis of successive impact times (delay coordinates) and much of the qualitative nature of the dynamics extracted. This may be very useful in the context of limited access to continuous state variable measurement, although there is the possibility of important oscillatory behavior occurring between impacts (especially when $X_c < 0$).

Before leaving this chapter, we add a few comments. First, these types of systems with piecewise linear characteristics are relatively common in mechanical engineering as well as other branches of science. Second, piecewise linear maps have proved popular archetypes in theoretical studies, partly because of the relative ease of extracting closed-form analytic solutions (Chin et al., 1994). For example, it is often possible to

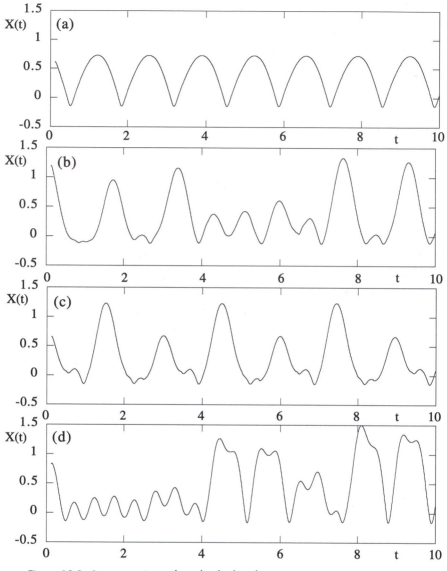

Figure 12.9: Some experimental results displayed as time series. $F=0.11$; $X_c=-0.1$; (a) $\Omega = 1.5$, (b) $\Omega = 2.35$, (c) $\Omega = 2.68$, (d) $\Omega = 3.8$.

obtain a complete response by matching solutions at the boundaries of linear subdomains (Shaw and Holmes, 1983). Third, since the occurrence of the nonlinearity is discrete, the abrupt onset of a qualitative change of behavior may produce behavior not encountered in smooth dynamical systems. In fact, low-velocity *grazing* bifurcations have been shown to underlie a whole range of subtle behavior (Nordmark, 1991; Nordmark, 1992; Begley and Virgin, 1998). This circumstance is more likely to occur in cases where the impact barrier is placed in a negative position such that a linear response loses its stability as it grazes (i.e., just comes into contact with) the discrete change in stiffness. Similar behavior has recently been observed in systems with Coulomb damping: Subtle transitions between stick and slip lead to *hesitations* (Begley and Virgin, 1997).

Chapter 13

Two-Frequency Excitation

13.1 Subharmonic Forcing

We now consider what happens to the Duffing oscillator when subject to a two-frequency base excitation of the form

$$x_t(t) = A_1 \sin(\omega_1 t + \phi_1) + A_2 \sin(\omega_2 t + \phi_2). \qquad (13.1)$$

This type of forcing, which leads to combination tones (subharmonic) and almost-periodic behavior, has been the subject of a number of classic analytic studies including work by Duffing (1918) and Helmholtz (Rayleigh, 1945), who considered a quadratic oscillator exhibiting *difference tones* and its relation with human hearing (the cochlea is nonlinear). The approximate analytical techniques of harmonic balance and perturbation theory, and indeed many of the stability approaches (e.g., computation of CMs), introduced earlier in this book can also be applied, with some care, to this type of problem (Stoker, 1992). Again, structural mechanics provides a useful context with combination resonances a focus of attention (Plaut, HaQuang, and Mook, 1986; Plaut, Gentry, and Mook, 1990).

Since the experimental–theoretical correlation is well established by now, the material presented in this chapter will be almost entirely based on experimental data. Suffice it to say that numerical integration of the underlying equations of motion confirms the experimental results.

The key to determining the anticipated form of response is whether the frequencies are commensurate or not. In this section we will consider the case where there is a rational multiple between ω_1 and ω_2 (this ratio is known as the rotation number). The system now evolves in a four-dimensional phase space with position, velocity, and the two forcing phases effectively providing the unique location at a given instant of time, that is,

$$(x, y, \theta_1, \theta_2) \in \mathbb{R}^2 \times S^1 \times S^1, \tag{13.2}$$

where $y = \dot{x}$, $\dot{\theta}_1 = \omega_1$, and $\dot{\theta}_2 = \omega_2$. The case where the two driving frequencies are incommensurate will be considered in the next section, in which case the output will then be nonperiodic (in fact typically quasi-periodic).

Given the large number of parameters now available, we shall just consider two frequency ratios, one rational and the other not, with the first frequency component held fixed throughout. The forcing phases will be set to zero in Equation (13.1), and the forcing amplitudes will be set equal with three different values used to illustrate a typical range of behavior.

Initially, we impart a (symmetric) base motion of the form

$$Y_t(t) = -0.075 \left[\sin(0.79(2\pi)t) + \sin(0.395(2\pi)t) \right], \tag{13.3}$$

where the amplitude is given in terms of track units and the frequencies (in the ratio 2:1) are normalized with respect to the natural frequency of the cart ($\omega_n = 1.26$ Hz). The first (dimensional) frequency was chosen as 1 Hz in order to stimulate resonant-like motion. Based on our knowledge of the frequency response of linear systems, we expect that despite the equality of the two forcing amplitudes in Equation (13.3), the effect of the second (lower frequency) component on the transmitted force is relatively minor (see Figure 2.6(b)). The oscillator acts as a filter to attenuate somewhat the lower frequency component since it is further from resonance.

The input was achieved by placing the cart–track assembly directly onto the shake table described in Section 6.3. Figure 13.1 shows the input in terms of (a) a directly measured time series, (b) the position of the shake table plotted against itself one-quarter cycle later, and (c) a frequency spectrum of the time series (with frequency in Hz). The curve in Figure 13.1(b) is a form of Lissajous figure (Beckwith et al.,

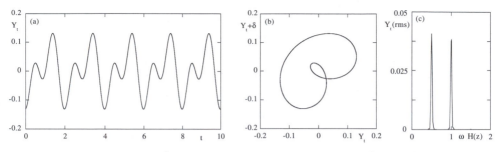

Figure 13.1: A period-two input (Equation 13.4). (a) Time series; (b) phase-lag projection; (c) frequency spectrum.

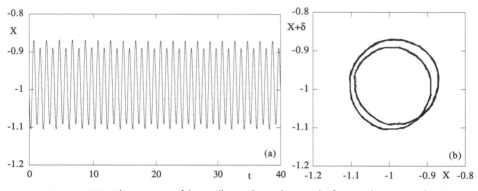

Figure 13.2: The response of the oscillator when subject to the forcing characterized in Figure 13.1. (a) Time series; (b) pseudo-phase projection. $\Omega_1 = 0.79; \Omega_2 = 0.5\Omega_1; F = 0.075$.

1993) and, because the data are projected down the time axis, it gives an effective confirmation of the repetitive nature of the input signal. The data on which the power spectrum is based were subject to a Hanning window and a bandpass filter $(0.1 - 3$ Hz$)$. Here, the vertical axis is linear and plotted in terms of root mean squared values of the shake table position.

The forcing (and hence amplitude of the cart) is sufficiently low that the response (Figure 13.2) is essentially linear with the expected subharmonic of order two contained within the left-hand well. Again a mirror-image response is present in the right-hand well. Plotting the response against itself one quarter of a cycle later gives the pseudo phase projection shown in Figure 13.2(b). The sampling rate in the data acquisition was set at exactly 1/100th of the (first) forcing period. In

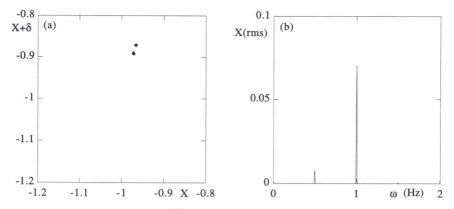

Figure 13.3: (a) Poincaré section; (b) frequency spectrum. Parameters the same as in Figure 13.2.

this way every 100th data point represented an appropriate Poincaré sampling, and since the motion in this case is periodic ($n = 2$), the resulting Poincaré mapping appears as shown in Figure 13.3(a). This form of Poincaré sampling is not as effective as the external triggering used elsewhere in this book owing to possible difficulties with sychronization between the sampling rate and forcing period over long durations. The frequency spectrum, with two distinct spikes at the forcing frequencies, shown in Figure 13.3(b), completes the summary.

This is essentially a linear feature, where a two-frequency output from a two-frequency input is expected. The more interesting case will be found when the forcing frequencies are incommensurate and the motion has a large amplitude. This will be considered in the next section.

13.2 Quasi-Periodic Forcing

One form of nonperiodic behavior found quite naturally in nonlinear systems is quasi-periodicity (Bergé, Pomeau, and Vidal, 1984; Nayfeh and Balachandran, 1995). This happens when more than one characteristic frequency is present in a system where the frequencies are *incommensurate*. Commensurate frequencies lead to subharmonics (or superharmonics): We saw in the previous section how a system subject to a two-frequency input responded with a two-frequency output using

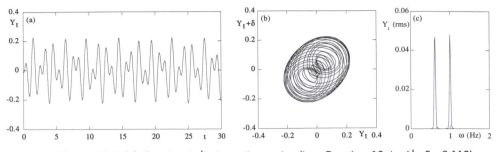

Figure 13.4: (a) Quasi-periodic input time series (i.e., Equation 13.4 with $F=0.112$); (b) corresponding phase-lag projection; (c) frequency spectrum.

the example of a 2:1 frequency ratio. We will now briefly consider the response of the system to a quasi-periodic input (Wiggins, 1987). In the absence of external periodic forcing, quasi-periodicity is also observed in certain flow-induced vibration problems in autonomous dynamical systems.

Now let's impart a base motion of the form

$$Y_t(t) = F [\sin(0.79 (2\pi)t) + \sin(0.488 (2\pi)t)], \qquad (13.4)$$

where the irrational frequency ratio was chosen on the basis of the Golden mean (Shenker, 1982; Ott, 1993), that is, $R = (\sqrt{5} - 1)/2$, and hence $\Omega_2/\Omega_1 = 1/R = 0.618$. Let's first consider the ensuing motion when the forcing magnitude is $F = 0.112$ (track units), such that the response is contained within a single potential energy well.

13.2.1 Single-Well Motion

With incommensurate frequencies, a typical shake table input now has the form shown in Figure 13.4, in which the nonrepeating nature of the excitation is evident from the measured position time series in part (a) and the phase-lag reconstruction (giving a complex-looking Lissajous figure) in part (b). However, we still have just the two frequencies as shown in the frequency spectrum (part (c)).

A typical (relatively small-amplitude) response to this quasi-periodic forcing is shown in Figure 13.5. Figure 13.5(a) shows a time series with a dominant frequency characteristic around 1 Hz but with some

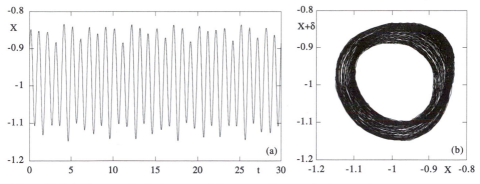

Figure 13.5: (a) Quasi-periodic time series; (b) corresponding pseudo-phase projection. $\Omega_1 = 0.79; \Omega_2 = R\Omega_1; F=0.112$.

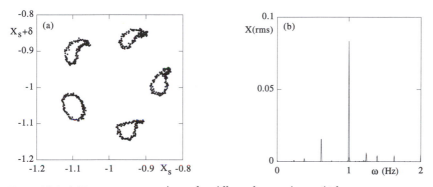

Figure 13.6: (a) Poincaré section taken at five different forcing phases; (b) frequency spectrum. The parameters are the same as in Figure 13.5.

apparent randomness superimposed. A pseudo-phase projection in Figure 13.5(b) gives a clearer indication of the undulations about periodic behavior.

Extracting a Poincaré mapping (Figure 13.6(a)) reveals an interesting new feature. Rather than producing a finite number of distinct points or strange attractor, the section consists of a closed curve. The trajectory was sampled five times during the forcing period (based on the first frequency), and hence five closed curves appear. The motion can be thought of as living on the surface of a torus – a time-invariant manifold (within a four-dimensional phase space), as shown schematically in Figure 13.7. During the evolution of the trajectory, the path completely fills up the surface of the torus. The two phase angles

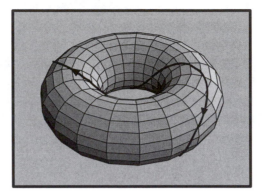

Figure 13.7: Schematic of the two-frequency motion winding around the surface of a torus.

(associated with the two frequencies and indicated by the contour lines) thus mark progress through the two ellipses, the product of which makes up the torus. Hence, the trajectory penetration through the surface of section appears as a closed curve. The five sections in Figure 13.6 help to visualize the evolution of the motion as the trajectory winds around the surface of the torus. This is also true for the case with commensurate frequencies (Figure 13.3), but in that case, the trajectory passes around the surface of the torus *exactly* twice before closing upon itself. If the frequencies are relatively close, the familiar beating effect is possible (Thomson, 1981).

In terms of predictability, the response shown in Figure 13.5 is quite different from the chaotic behavior observed earlier: This behavior is *not* sensitive to initial conditions and the main features (e.g., phase and amplitude information) can be obtained from appropriate trigonometric relations and superposition. Despite the apparent complexity of a quasi-periodic time series, the frequency spectrum, shown in Figure 13.6(b), gives two discrete spikes, which will again contrast with the more subtle complexity to be considered next.

13.2.2 Double-Well Motion

We now turn to the two-frequency form of excitation where the frequencies are again incommensurate, but the amplitude of forcing is sufficient to cause cross-well motion. We have seen how, provided the motion

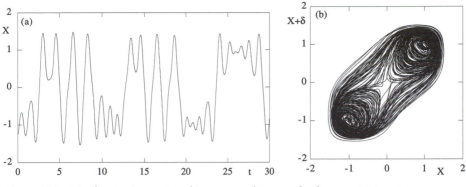

Figure 13.8: (a) Chaotic time series; (b) corresponding pseudo-phase projection. $\Omega_1 = 0.79; \Omega_2 = R\Omega_1; F = 0.141$.

was relatively low amplitude, the response was subharmonic or quasi-periodic depending on whether the frequencies were commensurate or not. In the event of large forcing amplitudes, it is perhaps not surprising to find that two-frequency excitation can also lead to chaos, and we conclude by making a brief study of just such a case, where a couple of new nonlinear features appear.

Increasing the forcing magnitudes from $F = 0.112$ (as used to generate the results in the previous section) to $F = 0.141$, we encounter double-well motion. A typical time series is shown in Figure 13.8(a), a response not dissimilar to that shown in Figure 9.1. The corresponding phase projection is shown in Figure 13.8(b), where many more cycles of motion have been plotted.

It is interesting to take a closer look at this time series to observe some interesting behavior also present in the Duffing oscillator with single frequency excitation. Figure 13.9 shows a couple of close-up views of the cart position in which the hilltop (indicated by the dashed line) is actually (briefly) crossed before the forcing pulls the cart back. This possibility contrasts with the unforced results shown in Figure 7.6. Note the lengthening of the period of motion during these two cycles, which does remind us of the unforced behavior. It is also possible for the cart to spend an extended period of time (perhaps 2 or 3 seconds) oscillating about the hilltop. There is, of course, an unstable periodic orbit that can be picked up for a brief duration before the CM with magnitude greater than 1 draws it away.

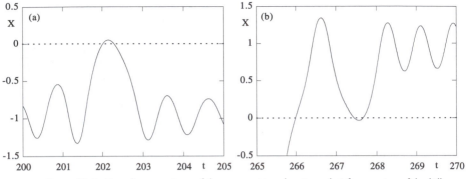

Figure 13.9: Two close-up views of the cart position showing a brief traversing of the hilltop.

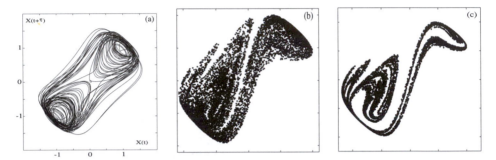

Figure 13.10: Numerical results based on the same nominal conditions as that which produced Figure 13.8. (a) A phase projection based on a reconstruction with a quarter-cycle delay; (b) A single Poincaré section; (c) A double Poincaré section. The scale is enlarged in parts (b) and (c) for clarity, with r = 1 corresponding to X_{max}.

13.3 Characterization of Chaos

We can also apply the techniques of establishing invariant measures of chaos evolving in a four dimensional phase space. Since some of these measures require a considerable amount of data and are somewhat sensitive to noise we base most of this section on numerical integration of the simplified Duffing system, that is, equation 5.38 with $\alpha = \mu = 0$, and assuming $\zeta = 0.025$.

The numerical equivalent to Figure 13.8(b) is shown in Figure 13.10(a) again using a quarter-cycle delay as a pseudo-velocity. Note that the omission of some of the nonlinear terms results in a more angular

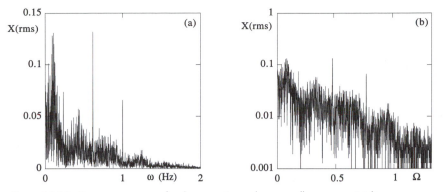

Figure 13.11: Frequency spectra for the experimental cross-well response, (a) linear axes, (b) semi-log axes. Parameters same as Figure 13.8.

phase projection, but again the qualitative correspondence is good. Conducting a standard Poincaré section of this response (triggered by the first forcing term) as shown in Figure 13.10(b), where only a subset of the computed points are actually plotted, gives a scattering of points with little apparent structure, at least in comparison with the single-frequency chaos (see Figure 9.4(b)). The Poincaré sections data have been scaled such that the maximum data point in each plot equals unity (to facilitate the dimension computations to be described later). However, an interesting means of observing dynamic evolution in a four-dimensional system in which a chaotic trajectory tends to fill up a large extent of the phase space is based on the concept of a double Poincaré section. Further details of this approach can be found in (Moon and Holmes, 1985), but the basic idea is that following a conventional Poincaré sampling, a second Poincaré section is triggered (based on proximity to the other frequency component of the signal). This is shown in Figure 13.10(c), where the second sampling, or *slice* (pulse width) corresponds to about 1/150th of the second phase angle. The fine structure of the attractor is now revealed. However, this approach (also known as Lorenz cross sections (Lorenz, 1984)) requires considerable data and would be prohibitively slow applied to the experimental system, especially when used as the basis for dimension calculations to be described shortly.

The frequency spectrum based on the *experimental* cross-well motion is shown in Figure 13.11(a) in which the broadband nature of the

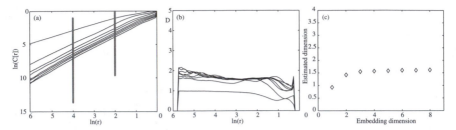

Figure 13.12: Dimension calculations based on the double Poincaré section (Figure 13.10(c)). (a) Scaling of the number (probability) of points falling within a small distance of a given location; (b) Computed slopes illustrating the plateau; (c) Estimated dimension based on the convergence to a specific slope as a function of embedded dimension.

frequency distribution is apparent. Part (b) shows the same data plotted on a logarithmic scale with the frequency in nondimensional units. Note the similarity with Figure 9.3, although for that single frequency excitation case, the power spectra were averaged to give the smoother appearance. Also, in contrast to the earlier frequency spectra in this chapter, Figure 13.11 was not based on any filtering. The equivalent result corresponding to the numerical data is essentially the same (with less noise) and not reproduced here.

A set of dimension computations (see section 9.5) will be described next. Recall that the motion on the surface of the torus described earlier was the result of attraction onto a lower-order manifold, and also that taking a Poincaré section effectively reduced the order of the attractor by one. A standard correlation dimension was conducted on the data assuming that the source of the data (producing the discrete time series) was unknown, e.g., the typical situation in an experiment. The first step is to reconstruct the attractor using delay coordinates, that is, the data is repeatedly embedded using time-delayed versions of itself until convergence (and hence topological equivalence) was achieved. Figure 13.12(a) shows the results of evaluating the pointwise correlation integral (see ref. (Tel, 1990)) based on the output of the double Poincaré section (Figure 13.10(c)). We observe that as the embedding dimension is increased (moving down) a constant slope is approached, thus giving an estimate of the correlation dimension (see equation 9.2). Here, natural logs are used, rather than the base 10 used in Chapter 9 or base 2 often preferred in electronics. The scaling region occurs over an intermediate range indicated by the vertical gray lines and Figure 13.12(b) (based on a least squares fit) reveals the dimension converging to a value of about 1.6, and also plotted as a function of embedding dimension in Figure 13.12(c).

A great deal of research has been conducted in to the extraction of meaningful dimension estimates, a problem particularly acute for large-order noisy systems. Even in this relatively simple system with low noise and access to large amounts of data there are a number of choices to be made. We have already commented on the choice of time-lag and number of dimensions in the embedding (Kennel et al., 1992), together with the corrupting effects of noise (Casdagli et al., 1991). The other subtleties include the amount of data needed (Theiler, 1990; Nerenberg and Essex, 1990), elimination of temporally correlated points (Theiler, 1986), the choice of scaling region (extent of the plateau) (Ding et al., 1993), effective use of subsets of data (Cusumano, 1990), etc. More sophisticated techniques include the use of singular value decomposition (SVD) to reduce noise (Broomhead and King, 1986) and mutual information to establish an appropriate delay for the embedding (Frazer and Swinney, 1986; Nichols, 1999).

The dimension estimate of approximately 1.6 was supported by independent estimates of 2.6 for the single Poincaré section (see Figure 13.10(b)) and 3.6 for the continuous time series data (see Figure 13.10(a)). With appropriate care for gross geometric effects the dimension algorithms can also be used to estimate the two-dimensional nature of the toroidal surface in the quasi-periodic case described earlier. It is interesting to note that equivalent results based on a double-well chaotic motion generated from frequencies related by a rational multiple revealed a dimension similar to (but slightly higher than) the single frequency case, where the taking of a second Poincaré section is unnecessary. This does, of course, raise some interesting questions regarding the practical distinction between commensurate and incommensurate frequencies. In this chapter we have focused on perhaps the two extremes: a simple (2:1) ratio and the irrational golden mean. A Lyapunov exponent computation based on the continuous trajectory resulted in the values $(0, 0, -0.153, +0.102)$, their (negative) sum satisfying the divergence theorem based on the damping value of $2\zeta = 0.05$ used in the simulations, and again confirming the link between LEs and dimension via the Kaplan-Yorke conjecture ($D_L \approx$ 3.6). Some analysis based on an extension of Melnikov theory suggests that additional forcing terms tend to increase the likelihood of chaos (Wiggins, 1987).

However, in this two-frequency case, the Poincaré sampling reduces the order of the system from four to three and then projects the results onto the plane. In order to reveal some further details of this behavior, together with a providing brief sketch of the route to chaos, we conclude this chapter by considering some additional aspects of quasi-periodicity.

13.4 Additional Features

Although a quasi-periodic output resulting from a quasi-periodic input, as described earlier, should not come as any great surprise, it is the occurrence of quasi-periodicity in a more general context in nonlinear systems (and especially in Hamiltonian systems) that presents considerable interest. Furthermore, extending these ideas to include systems with more than two frequencies leads to certain universal behavior (Ott, 1993). In a sense, the material presented in this chapter has only scratched the surface of dynamics in four-dimensional space. This is partly due to the constraint provided by a fixed frequency ratio.

The study of quasi-periodic behavior occupies quite an important role in the development of nonlinear dynamics. There are a number of reasons for this. The analytical treatment of undamped nonlinear oscillators subject to quasi-periodic excitation may lead to the difficulty of small divisors – a problem familiar to Poincaré (Stoker, 1992). This was also considered by Arnold (1965), who investigated the development of the circle map as a prototypical nonlinear system, finding some very interesting properties and rich bifurcation sequences (including the features of mode locking, Arnold tongues, and the Devil's staircase).

Specifically, one of the classic routes to chaos occurs as a result of torus breakdown. In this case a typical scenario in an autonomous system might be the following: A system in equilibrium loses its stability via a (supercritical) Hopf bifurcation resulting in a limit cycle (i.e., a pair of complex conjugate CEs leaves the negative real half-plane). This is followed by a secondary Hopf bifurcation in which the limit cycle now loses stability, resulting in quasi-periodicity as another frequency enters the picture (i.e., a pair of complex conjugate CMs leaves the unit circle). This is sometimes followed by torus breakdown, that is, the surface of the torus (and hence the closed curve of the associated Poincaré map) distorts, wrinkles, becomes fractal, and finally breaks up as the control parameter is gradually changed (Virgin, Dowell, and Conner, 1999), resulting in chaos.

The pioneering work of Ruelle and Takens (Ruelle and Takens, 1971) suggested the importance of quasi-periodicity (specifically with hyper-tori greater than degree 2) on the stability and, hence, observation of certain complex behavior. This had special relevance in fluids, suggesting a deterministic route in the transition from laminar flow to turbulence and providing some important experimental motivation (Swinney and Gollub, 1978; Brandstater and Swinney, 1987).

Clearly the power spectrum plays an important role in identifying qualitative changes in behavior in systems involving quasi-periodicity. Also, LE computations can be useful here, especially the Lyapunov spectrum (plotting LEs as a function of control parameter), where the two-periodic torus leads to two zero LEs with the other two constrained by the divergence theorem and reflecting the transition to chaos. Finally, we also note that quasi-periodicity can lead to strange nonchaotic attractors (Romeiras and Ott, 1987). In this case, although the structure of the attractor has fractal properties, none of the LEs of the system are positive. Strange nonchaotic attractors are also characterized by an interesting scaling of peaks in their spectral distribution.

Chapter 14

Global Issues

14.1 Introduction

In earlier chapters, we saw the means by which motion, bounded initially within a potential energy well, might spill over or escape either to infinity or to an adjacent energy well. In this chapter we take a closer look at global issues. We will see how basin boundaries and unstable fixed points have a considerable influence on behavior in the large. Dependence on initial conditions has been encountered earlier in this book in terms of multiple (point and periodic) attractors and the extreme sensitivity of chaos. We shall see that extreme sensitivity to initial conditions may also appear when the boundaries separating domains of attraction become fractal, causing transients to have arbitrarily long lengths (Eschenazi, Solari, and Gilmore, 1989; Grebogi, Ott, and Yorke, 1987; Gwinn and Westervelt, 1986). This is often a precursor of steady-state chaos. One specific aspect of interest is the appearance of indeterminate bifurcations. For the purposes of illustrating this behavior, we will revert back to the double-well Duffing oscillator of earlier chapters of this book. A chronological note here is that the experimental results to follow were obtained a couple of years after those described in Chapters 8 and 9 and, hence, some small adjustments appear in the basic system coefficients (Todd and Virgin, 1997b).

14.2 Dependence on Initial Conditions

One of the fundamental differences between a linear and a nonlinear system is that nonlinear systems often possess multiple stable solutions, and, hence, the final solution depends to an extent on the starting conditions. The standard theory of linear vibrations, even for high-order systems, obviates the need to consider this, with unique solutions capturing all possible initial conditions. We have seen that nonlinear systems (even unforced problems) typically have a variety of long-term solutions for a fixed set of parameter values. Although it can be argued that persistent (stable) solutions perhaps have the most practical importance (certainly in relation to their local region of phase space), it is the *unstable* solutions that have a profound influence on global behavior (Grebogi, Ott, and Yorke, 1986b).

Clearly, unstable solutions and the prescription of initial conditions are much more difficult to investigate from an experimental perspective (Virgin et al., 1998). The double-well potential system again provides a sound test bed on which to gain some initial insight into global behavior. We shall focus on the transition from smooth to fractal basin boundaries and discuss one or two implications that this has for predictability (Todd, 1996).

Consider, then, the equation of motion based on the full system equation (5.38). When the parameters are fixed at $\alpha = 0.81$, $\zeta = 0.002$, $\mu = 0.02$, $F = 0.03$, and $\Omega = 0.89$, four period-one attractors coexist in the vicinity of primary resonance: resonant and nonresonant oscillations in each potential energy well. These periodic responses are quite close to being harmonic, although the resonant oscillation has a noticeable egg-shaped phase trajectory (see Figure 8.4(b)). Figure 14.1 shows a conventional initial condition study in which the equation of motion is integrated from a fine grid of initial conditions (totalling 90,000 simulations). After allowing 100 cycles of motion, the resulting steady state (in terms of a Poincaré section defined by a constant value of the forcing phase) is identified and shaded. The shading scheme and subsequent notation is as follows:

- white – (S_n^-), the nonresonant (small-amplitude) oscillation in the left-hand well;

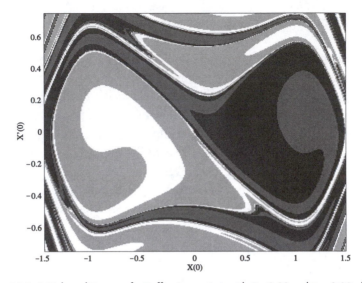

Figure 14.1: Initial condition map for Duffing's equation, with $F = 0.03$ and $\Omega = 0.89$, showing smooth basin boundaries based on numerical integration from a regular grid of starts.

- light gray – (S_r^-), the resonant (large-amplitude) oscillation in the left-hand well;
- medium gray – (S_n^+), the nonresonant (small-amplitude) oscillation in the right-hand well;
- dark gray – (S_r^+), the resonant (large-amplitude) oscillation in the right-hand well.

The structure is quite fine, especially for high initial velocity where the bands are tightly intertwined. However, the boundaries (separatrices) between the domains of attraction are smooth. A close-up view is provided in Figure 14.2. We see that a rough lack of precision in the initial conditions may lead to a degree of uncertainty regarding the final outcome for a given trajectory. However, this is not the *extreme* sensitivity to be encountered later, and we also note at this point that this, and subsequent results, relate to a deterministic situation and thus (in principle) are perfectly repeatable. An efficient computational technique has also been developed based on cell-to-cell mapping (Hsu, 1987; Tongue, 1987).

In earlier studies, the technique of time-lag embedding (see Section 6.5) was used to faithfully reconstruct phase trajectories based on

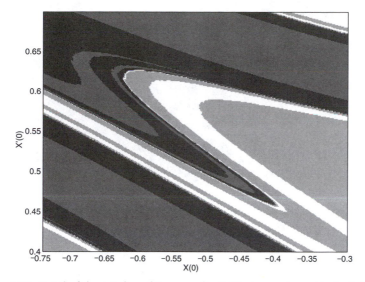

Figure 14.2: Detail of the initial condition map for Duffing's equation, with $F = 0.03$ and $\Omega = 0.89$.

using only a subset of measurement state variables. This was shown to have particular benefit when conducting experiments. Therefore, to exploit time-lag embedding in terms of *initial* conditions, account must be taken of the fact that rather than specifying a position and velocity at a certain time (e.g., $t = 0$) we need two positions on a single trajectory separated by a fixed (delayed) time interval, δ. In numerical terms, this can be achieved by employing a *shooting* method to solve the boundary value problem: An estimate is made of the initial velocity, the initial value problem is solved and then compared with where it should be a specified later time, and a Newton–Raphson iterative procedure is used to drive this error toward zero. This whole process is then repeated over the initial condition space. Figure 14.3 shows the results for the same set of parameter values as used in Figure 14.1. In these figures δ is again taken as one quarter of the forcing period. Clearly there is a good topological equivalence with the diagonal (bottom left to top right) in this projection representing states initially at "rest." A very small number of nonconverged points do appear, but their number can be minimized by appropriate choices in the initial guesses akin to path-following routines (Kaas-Petersen, 1987; Seydel, 1991).

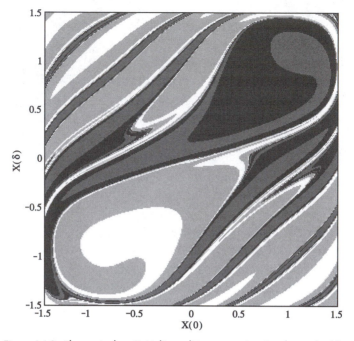

Figure 14.3: The equivalent "initial" condition map using time-lag embedding.

14.2.1 Stochastic Interrogation

In the above studies, it was a relatively straightforward task to simply specify the initial position and velocity or time-delayed starting points in the numerical algorithm. In an experimental context, this is not so easy. However, a technique called stochastic interrogation, developed by Cusumano and his coworkers (Cusumano and Kimble, 1995), can be used to produce a set of randomly distributed initial conditions by employing a scheme of stochastic disturbances to various steady-state solutions. This is clearly related to the method of using perturbation-induced transients for unstable fixed point characterization, detailed in Chapter 9, and is also used to determine the deteriorating stability of a stable fixed point in Chapter 11. Now, the various fixed points will be subjected to relatively large disturbances, and the resulting transient motion will be tracked (on the surface of section) until convergence to a steady state is achieved.

Consider the time series shown in Figure 14.4. This is actually based on an extended experimental time series during which the cart

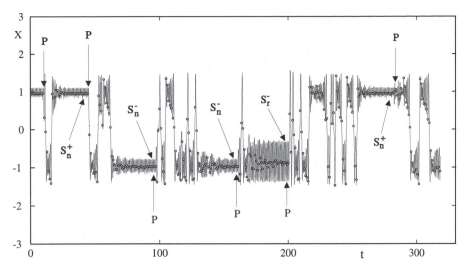

Figure 14.4: A randomly perturbed time series showing the implementation of the stochastic interrogation procedure. A "P" indicates a perturbation; the sub- and superscripts on S are as listed in the text.

was randomly perturbed from, and between, the four coexisting period-one attractors. The perturbation was achieved by imposing a short burst of randomly chosen forcing conditions before switching back to the base set of parameters. By carefully noting the initial transition through the surface of section, subsequent penetrations are tracked, and when the transient finally settles onto a steady-state solution, the initial conditions (and their corresponding images) are attached (shaded) to their final outcomes. This whole procedure is then automated (and based in principle on a single run) and a large collection of initial states can thus be labeled. There are a number of issues concerning the scheduling of perturbations with white noise providing perhaps ideal, but difficult to achieve, disturbances. They must be large enough that the subsequent trajectory actually leaves the local basin of attraction (at least some of the time), and they must be sufficiently random to enable the entire region of the initial condition space under consideration to be visited. Sometimes the inherent time scale of the dynamics (due to the inertia of the mechanical system) prevents certain regions of the phase space being visited much prior to the first penetration of the surface of section (Virgin et al., 1998).

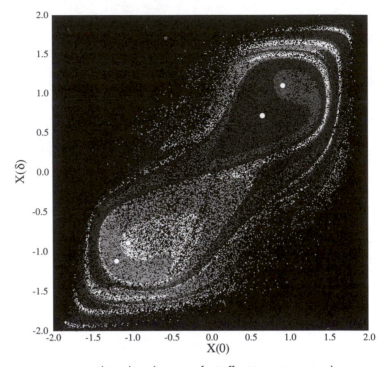

Figure 14.5: A numerical initial condition map for Duffing's equation, using the same parameter values as in Figure 14.3 but generated using stochastic interrogation.

First, this approach will be used on the numerical model as a test. Figure 14.5 shows the results of initiating 50,000 randomly distributed starting conditions. Also shown in this figure are the fixed points. The concentration of points toward the diagonal reflects the relative ease of generating initial conditions induced by a relatively low velocity start (i.e., a higher probability).

This technique is now applied to the experimental system. Figure 14.6 shows the result of about 3,000 points generated in the manner of Figure 14.4. The random sequence of perturbations to achieve this figure was generated within the LabVIEW software package. Clearly, there are a number of alternative ways of disturbing the system from its steady state. In the hardening spring system of Chapter 11, perturbations were achieved using a braking mechanism. For the cart system, the forcing parameters were subject to a short burst of stochasticity by making step changes to the forcing amplitude and frequency (and hence forcing

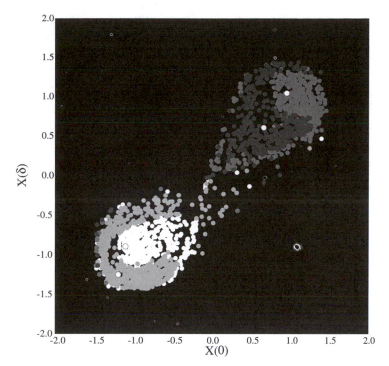

Figure 14.6: An experimental initial condition map for Duffing's equation.

phases) at randomly chosen times. This new set of forcing conditions was then shut off with the system returning to its base forcing parameters. The first subsequent penetration of the surface of section (and subsequently a quarter of a cycle later) defined the *initial conditions*, which were then followed until the trajectory settled back to the original, or some other, attractor.

The equivalence with the numerically generated picture is close. The scarcity of points near the origin is partly associated with the fact that transients typically do not spend much time in the vicinity of the hilltop, and it is also relatively difficult to perturb the system to this region of the phase space. The slight overlap of the data points in the experimental plot is purely a graphical artifact (with smaller points the necessarily limited amount of data would not produce the contiguous shades and would, therefore, be difficult to assess). This technique will also be illustrated in the appendix when applied to an electric circuit.

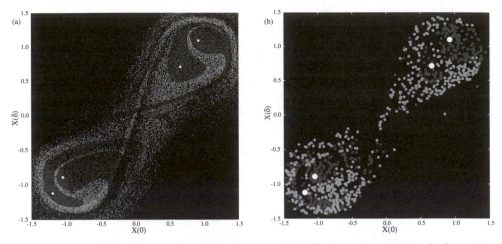

Figure 14.7: A numerical initial condition map for Duffing's equation, showing the fast and slow initial conditions. (a) Numerical; (b) experimental.

14.2.2 Transient Lengths

There is another piece of information that is also generated by the stochastic interrogation procedure: the time taken to achieve convergence. Figure 14.7(a) divides the initial conditions into two shades of gray: dark – those that converge quite rapidly (i.e., within 10 forcing cycles) to any of the fixed points, and light – those that take longer to settle (Pezeshki and Dowell, 1987). The equivalent picture generated from the experiment is shown in Figure 14.7(b). The "cutoff" between fast and slow in these two figures is the same. This division is somewhat arbitrary, but we note that owing to experimental noise a different tolerance must be used to satisfy convergence, and again, there is an inevitable discrepancy due to damping. As expected, motions arising from initial conditions close to the locations of the fixed points quickly settle back down to the same steady state. However, we also begin to see some more global effects as the fingers of the *slow* initial conditions extend around and beyond the remote fixed points (including those in the other potential energy well). Clearly, this result is associated with the stable manifolds of the unstable fixed points (which lie between the stable fixed points in each well). There are also some initial conditions that take a very long time to settle. In cases where the basin boundaries are fractal, these transient lengths

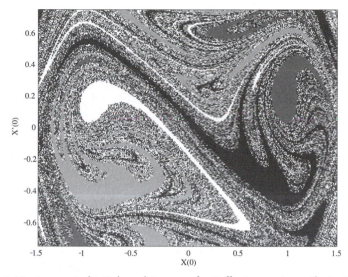

Figure 14.8: A numerical initial condition map for Duffing's equation, with $F = 0.09$ and $\Omega = 0.795$.

may be infinitely long and have been labeled as transient chaos (Ott, 1993). A typical time series was shown in Figure 9.12. Another aspect of very long transients is the occurrence of *dwelling* associated with bottlenecking and ghost solutions. This aspect will be revisited in Appendix A.

If the forcing parameters are changed to $F = 0.09$ and $\Omega = 0.795$, fractal basin boundaries are encountered (Mandelbrot, 1983; Moon, 1985). Figure 14.8 shows the conventional, numerically integrated initial condition map. It can be compared with its stochastically interrogated, time-delayed equivalent in Figure 14.9. In this case the fixed points are still attractors, that is, they are surrounded by domains of attraction, but the boundaries separating regions of initial conditions are by no means smooth. These intrusions into the resonant attractor basins occur rather quickly and lead to a rapid loss of basin integrity. Clearly this effect would not be detected by a local analysis. This has clear consequences for predictability based on any initial condition or parameter imprecision.

Figure 14.10 shows the results based on the experimental system. Partly because of the relative scarcity of data points, the fractal nature of the basin boundaries is not so easy to discern. Each fixed point is

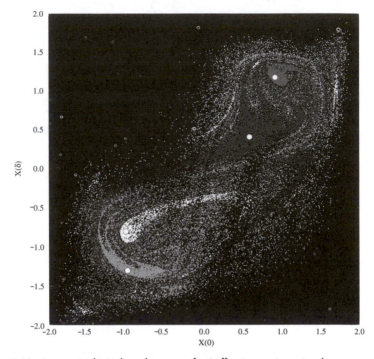

Figure 14.9: A numerical initial condition map for Duffing's equation using the same parameters as in the previous figure but with time-lag embedding and randomly generated initial conditions.

surrounded by a solid shade, but the shades become intertwined and somewhat difficult to assess in this figure. Some of the data points took a very long time to settle onto an attractor (Todd, 1996), and we note that this set of parameter values is close to that used to obtain the transient chaos in Figure 9.12.

On subsequent increase in the forcing amplitude the two resonant attractors disappear and initial conditions leading to the chaotic attractor described in Chapter 9 start to occur. Figure 14.11 shows the experimental basins for $F = 0.15$ and $\Omega = 0.74$. Owing to the limitations of the gray shading scheme this figure requires a little interpretation. Figure 14.11(a) shows those initial conditions (experimentally determined using stochastic interrogation) that lead to the nonresonant attractors (white for S_n^-, and gray for S_n^+). Figure 14.11(b) shows initial conditions that subsequently did not lead to either of the two small-amplitude attractors (and,

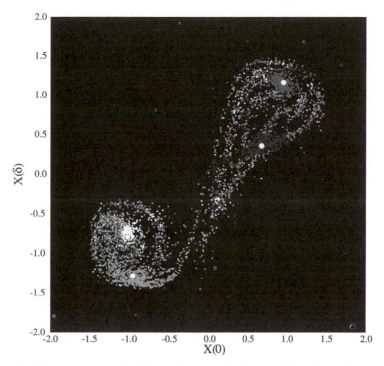

Figure 14.10: An experimental initial condition map for the case of fractal basin boundaries.

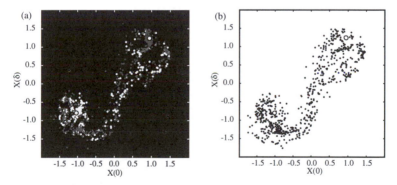

Figure 14.11: Experimental basins of attraction for the case of $F = 0.15$, $\Omega = 0.74$. (a) Initial conditions leading to the nonresonant periodic attractors; (b) initial conditions leading to other (generally chaotic) attractors.

in fact, typically led to chaos). Thus, these plots can be superimposed to reveal further reduction in the size of the (smooth) nonresonant basins but considerable mixing on the global level.

We shall now consider these results against the backdrop of global manifold interactions and discuss an analytic technique that has been used successfully to predict the transition between smooth and fractal basin boundaries.

14.3 Manifold Interaction

The double-well Duffing system provides a classic example of the possible interaction of stable and unstable manifolds to produce homoclinic tangling. The importance of this is based on the fact that under certain conditions, manifold entangling leads to the presence of horseshoe maps: an identified precursor of chaos caused by an extreme sensitivity to initial conditions (Guckenheimer and Holmes, 1983).

Without damping or external forcing, the underlying potential energy of the system possesses two stable minima and an unstable maximum. This hilltop (at the origin) forms the common point of two homoclinic orbits as described in Chapter 7. The addition of a small amount of forcing and damping turns these equilibria into their periodic counterparts with trajectories now living in their three-dimensional phase space. These periodic solutions were considered in Chapter 8. But again, in keeping with the general progression in this book, the interesting nonlinear events occur when the external forcing is increased further. We have already seen how eventually chaos and escape may occur, but in order to examine this scenario more fully, we need to observe the changes taking place in a global sense (Katz and Dowell, 1994).

Poincaré sectioning, again, usefully converts the three-dimensional flow into a discrete two-dimensional mapping. An unstable periodic orbit (a saddle cycle) appears as a single fixed (Poincaré) point, possessing both stable and unstable manifolds (invariant paths along which Poincaré points get mapped toward and away from the fixed point, respectively). For relatively low levels of forcing, these manifolds do not intersect (and, for example, the stable manifolds of the unstable fixed points form the smooth boundaries in Figure 14.1).

However, for increased levels of forcing, a homoclinic intersection occurs, and it can be shown that once the manifolds have intersected once, they, in fact, intersect an infinite number of times. This leads to severe tangling and complexity, behavior anticipated by Poincaré (Poincaré, 1921). The resulting strong folding and stretching of the phase space (in a kind of horseshoe geometry) thus results in an extreme sensitivity to initial conditions. This has profound implications for predictability.

14.4 Melnikov Theory

A relatively successful analytic contribution to global bifurcation is based on a perturbation approach to assess the proximity between manifolds. The Melnikov function is a measure of the separation between unstable and stable manifolds of the underlying Poincaré map with a homoclinic (or heteroclinic) orbit appearing in the unperturbed system (akin to Figure 7.4). Further details of this approach can be found in Refs. (Melnikov, 1963; Guckenheimer and Holmes, 1983; Ketema, 1992), including an example based on the twin-well Duffing system. A brief outline is included here. The Melnikov function is defined as (Moon, 1992)

$$M(t_0) = \int_{-\infty}^{\infty} g^* \cdot \nabla H(x^*, v^*) \, dt, \qquad (14.1)$$

where $g^* = g(x^*, v^*, t + t_0)$, and x^* and v^* are the unperturbed solutions of the homoclinic orbit. The underlying Hamiltonian, H, is used as the basis of a perturbation study, which then determines the simple zeros of the Melnikov function. Couched in terms of the parameters of the problem, critical values of forcing amplitude, say, can be obtained when transverse crossing occurs for a given forcing frequency. This criterion provides the onset of fractal basin boundaries. Steady-state chaos tends to occur for slightly higher forcing levels, and hence this criterion tends to provide a lower bound, or necessary condition, for chaos (Guckenheimer and Holmes, 1983).

Let's now assume that the standard form of Duffing's equation makes an adequate representation of the experiment (i.e., assume the nondimensional parameter α is small and neglect Coulomb friction). In this case the equation of motion becomes

$$X'' + 2\zeta X' - \frac{1}{2}X + \frac{1}{2}X^3 = F\Omega^2 \sin(\Omega\tau), \qquad (14.2)$$

and using a typical damping value of $\zeta = 0.025$, we can evaluate the Melnikov criterion as

$$F = F_c = \frac{8\zeta}{3\sqrt{2\pi}\Omega^3} \cosh(\pi\Omega/\sqrt{2}). \qquad (14.3)$$

This function can then be plotted in the forcing parameter space. However, because of the less than exact modeling associated with the level of approximation inherent in Equation (14.2) with reference to the experiment, we simply note that the parameters on which the initial condition plots earlier in this chapter were based give critical forcing amplitudes of $F \approx 0.08$ for $\Omega = 0.89$ and $F \approx 0.09$ for $\Omega = 0.795$. We see that the case with smooth boundaries falls beneath the critical level suggested by the Melnikov criterion, and the fractal basin boundary case occurs at just about the critical level.

A Melnikov analysis has also been conducted for the asymmetric escape equation (10.1), which, again, in terms of the forcing parameters, is given by (Thompson, 1989)

$$F = \frac{\beta \sinh(\pi\omega)}{5\pi\omega^2}. \qquad (14.4)$$

We note here that $\omega \equiv \Omega$ and $\beta = 2\zeta$, and since this expression was based on direct forcing, we can simply divide the right-hand side by ω^2 to obtain the transmissible excitation case. Now the forms of Equations (14.3) and (14.4) are very similar when mapped into the (F, ω) plane.

Equation (14.4) was found to be in close agreement with numerical studies on the onset of the homoclinic tangency, and (again incorporating the base excitation effects) when plotted in the plane of forcing parameters, this tangency occurs well below the escape regions described in Chapter 10. Indeed, at these values the periodic attractors are relatively unaffected by these fractal basin boundaries. However, further global events have a profound influence on the general behavior (and predictability) of this oscillator under relatively large levels of excitation (Todd and Virgin, 1997b).

14.5 Bifurcations in Control Space

Now armed with some insight into the global structure of the twin-well oscillator, we revisit some of the earlier results on local bifurcational behavior to try to piece together a more general picture of the behavior of Duffing's equation.

We have already encountered how the periodic solutions of the Duffing oscillator typically lose their stability via saddle-node (fold) and period-doubling (flip) bifurcations. These instabilities were projected onto the plane of (F, Ω) in Figure 8.3 based on approximate analytical methods. These bifurcations were also underlying parts of the quasi-static escape diagram (Figure 10.4), where a control parameter was varied quasi-statically. To gain a more complete picture and verify these bifurcations in experiments, Figure 14.12 summarizes the bifurcation plot in the plane of forcing parameters for the Duffing oscillator for both numerical data (a) and experimental simulation (b). The fold lines (which meet at a cusp) are the locus of resonant amplitude jumps. A threshold on the forcing amplitude can be observed, below which the response will be essentially unique and almost linear. The region of hysteresis can be seen to grow with forcing amplitude, with the jump up to resonance given by the upper fold line. The flip bifurcation typically requires a somewhat larger forcing to be observed.

The steady-state escape and transient escape curves (only the minimum escape results are shown here; the fingerlike incursions from Figure 10.7 still exist) are basically the same as mapped out in Chapter 10. It is slightly misleading to label these curves in this way since all escape is necessarily transient, and in the double-well system, a trajectory may very well return at a later time. The distinction is that the steady-state points are the result of a parameter being changed quasi-statically and, hence, allowing local bifurcations to evolve. Whereas, the transient points are all based on the outcome resulting from a motion started from rest (i.e., with significant transients induced). We see that, over a certain range of frequencies, the steady-state escape curve coincides with the fold line where the jump to resonance causes a sufficiently large excursion for the motion to traverse the hilltop. This mechanism was discussed in Chapter 10 with regard to the modified (single-well) track shape. Furthermore, we also see that the steady-state

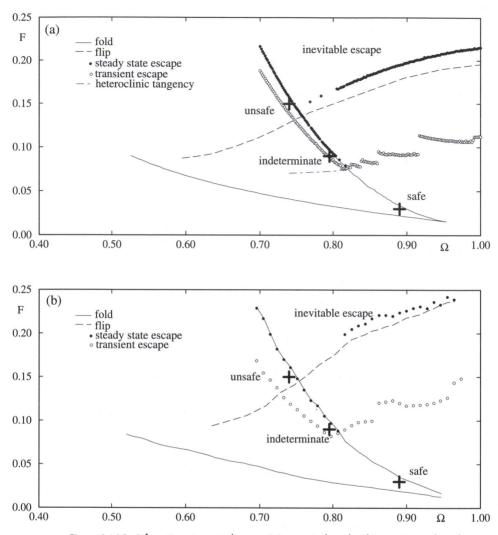

Figure 14.12: Bifurcations in control space: (a) numerical results, (b) experimental results.

escape curve appears just above the flip line for higher forcing frequencies. The mechanism in this range of parameters is that the flip initiates a sequence of period-doubling bifurcations (with slowly growing amplitude of response), which leads to chaos and is almost immediately followed by escape (see Chapter 10) and cross-well chaos. Further details of the transition from single-well to cross-well motion can be found in Ref. (Katz and Dowell, 1994). The triangular region, labeled "inevitable escape," represents those regions of parameter space where very strong forcing prevents the motion from being contained in a single well (for any initial conditions). We also note that for a given forcing amplitude, escape is more likely to occur in the vicinity of resonance, a result not unexpected given resonant amplification in linear systems. Before concentrating on some unexpected behavior close to the optimal escape conditions, we will briefly review some of the typical sequences of events as that occur a parameter is very slowly evolved.

Let's read off values from Figure 14.12(a) and follow some generic sequence of events. Suppose we fix Ω at 0.85 and gradually increase F from zero. Obviously, at zero forcing amplitude, the mass is not moving and equilibrium is a point attractor (notwithstanding the presence of dry friction). A small-amplitude period-one attractor grows just as if the system were a linear oscillator (but with increasing asymmetry). When the forcing magnitude reaches about 0.055, the system jumps up and after some transient motion converges onto the resonant branch. Note that the lower fold line (associated with the jump down) is not enountered in this sequence – remember that we are really viewing a three-dimensional picture with the response measure a projection out of the page. With increased forcing, we expect the appearance of fractal basin boundaries (Equation 14.3 suggests a forcing amplitude in the vicinity of $F = 0.082$). However, the oscillation is still relatively unaffected by this global event at this point. On further increase in forcing, a flip bifurcation is encountered (at about $F = 0.17$), leading to the growth of a period-two solution, cascading into chaos, and rapidly being followed by escape (at about $F = 0.19$) and cross-well chaos of the type described in Chapter 9. Now, if the forcing magnitude were reduced (still with the same fixed forcing frequency), the cross-well motion would gradually become captured by one of the energy wells, typically leading to a jump from resonance and, henceforth, to the stationary condition. At Ω equal to 1.0, the motion would simply grow in amplitude until encountering a flip at about $F = 0.2$ followed by the same escape scenario. A similar picture occurs if the system is started with a forcing amplitude fixed at, say, 0.16, and the forcing frequency is reduced (from above resonance) with a flip encountered at about $\Omega = 0.83$. Starting from a forcing amplitude of 0.05 also leads to a jump (up or down depending on which solution branch the motion is on), although no escape is found at this relatively small forcing amplitude. This last transition was seen in Figure 8.9.

Three crosses are marked on Figure 14.12 as "safe", where motion remains within the confines of the local potential energy well, "indeterminate", and "unsafe," where the motion spills out of the local potential energy well. These points are representative of their local regions of parameter space and their classification is determined by some of the global issues surrounding manifold intersections and fractal basin boundaries encountered earlier. The really interesting behavior occurs in the vicinity of intermediate forcing conditions where the postjump outcome may be indeterminate.

14.6 Indeterminate Outcomes

We notice that in Figure 14.12(a) the steady-state escape points appear multivalued over the parameter range $0.76 < \Omega < 0.8$, and this indecision was also noted when using a finer frequency increment in experiments. We also notice that part of the steady-state and transient escape curves are quite close to each other in this region. Suppose we have a steady-state oscillation with $F = 0.09$ and Ω very close but just smaller than the fold line in the vicinity of $\Omega \approx 0.79$. Any small increase in F, or perturbation, will typically cause a jump to resonance. This certainly causes a transient, but what subsequent motion will persist? Figure 14.13 shows eight experimental time series that were created under very similar conditions (the conditions of rest for the initial position at $+1$ are relatively easy to achieve in an experiment) but with slightly different forcing frequencies. The outcome is very sensitive to parameter values and three alternative outcomes are evident: restabilization back onto the small-amplitude (nonresonant) attractor; attraction to the large-amplitude (resonant) attractor; or escape from the well (to the left-hand well in this case). A similar lack of postjump predictability also occurs, given some imprecision in initial conditions or even sweep rate (see Section 8.5). The presence of certain global bifurcations leading to fractal basin boundaries thus makes the three possible outcomes of Figure 14.13 a more subtle, and more profound, situation than the two possible outcomes in each of Figures 10.3 and 10.9.

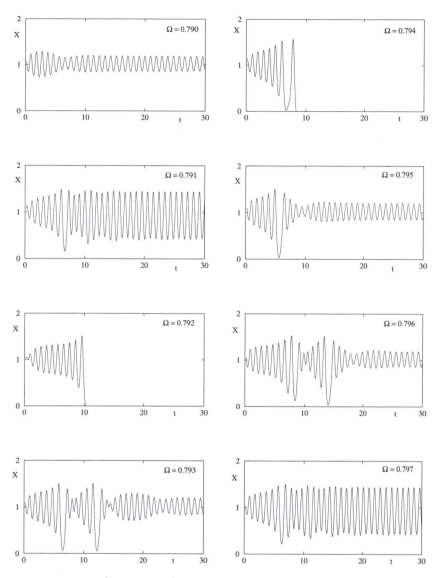

Figure 14.13: Sample experimental time series at $F = 0.09$ and $\phi = 0$ showing trajectories that undergo different long-term behavior under the influence of small perturbations in Ω. All the series were started from the same nominal rest position.

The events governing this situation are quite complicated with a sequence of homoclinic and heteroclinic bifurcations. More details can be found in Refs. (Soliman and Thompson, 1989; Thompson, 1992; Soliman and Thompson, 1992; Thompson and Soliman, 1990; Ueda et al., 1990; Soliman and Thompson, 1991; Stewart and Ueda, 1991; Soliman and Thompson, 1992; Nusse, Ott, and Yorke, 1995; Nusse and Yorke, 1996; Todd and Virgin, 1997b). The potentially important engineering implication for this behavior is the following: A strong familiarity with linear theory necessarily focuses stability issues on local behavior with a certain confidence in predictability. However, this last section shows us that although a system may be locally stable, its basin of attraction may shrink very rapidly, thus rendering the system susceptible to finite disturbances as well as removing our ability to predict certain behavior. An attempt to quantify this situation is based on the idea of global basin integrity curves, which seek to measure the probability of certain outcomes (Soliman and Thompson, 1989; Thompson and Soliman, 1990; Thompson, 1992). The small curve labeled "heteroclinic tangency" in Figure 14.12(a) is a key transition in this kind of indeterminacy and, in this study, was estimated on the basis of the *Dover cliff* effect: the rapid change in the size of the basin of attraction described in Ref. (Soliman and Thompson, 1992).

Clearly, this last set of results are a long way from the linear behavior encountered at the beginning of this book. It is tempting to say "a world away", but of course, since most systems in the physical world *are* nonlinear, to a certain extent, we should properly view linear systems as a relatively restricted subset of dynamics. There are also a host of other nonlinear (especially global) effects with profound implications, but illustrating these experimentally becomes an increasingly challenging task.

Appendix A

A Nonlinear Electric Circuit

A.1 Introduction

Apart from describing the behavior of the mechanical systems detailed in this book, nonlinear ordinary differential equations are, of course, used to model a large variety of physical systems. Electrical circuits with their lumped parameters can be very well modeled by differential equations. Within certain limits, these systems will be typically linear and methods based on the Laplace and Z-transform approaches incorporating transfer functions and control theory have proved to be extremely successful (Mees, 1981; Luenberger, 1979; Shahian and Hassul, 1993). However, there are circumstances in which significant nonlinearities cannot be ignored, and the example detailed in this appendix is a simple electrical circuit contrived to mimic Duffing's equation in much the same way as the mechanical analog (Trickey, 1997).

The compelling force–voltage analogy provides a strong correspondence between the electrical elements of resistance, capacitance, and inductance, and the mechanical elements of damping, stiffness, and mass. The time evolution of the state variables of charge and current map out trajectories and exhibit all the standard features (e.g., exponential free decay and forced resonance), observed earlier in this book. Thus, this appendix serves to illustrate the ubiquity of some of the nonlinear behavior described earlier. The high signal-to-noise ratio of typical circuits and

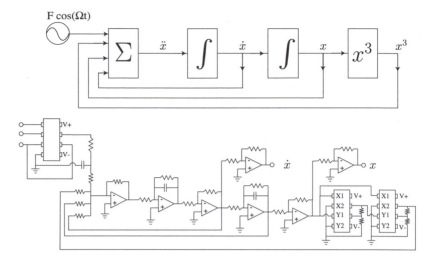

Figure A.1: Schematic of the electric circuit modeling Duffing's equation.

relatively fast characteristics will be exploited to confirm some generic nonlinear features and will highlight the interesting phenomenon of bottlenecking, in which a transient trajectory dwells in the vicinity of a ghost solution (Trickey and Virgin, 1998).

Ordinary differential equations are often represented in a block diagram form such as the top part of Figure A.1, which is a schematic representation of

$$\ddot{x} + 2\zeta\omega_n\dot{x} - (\omega_n^2/2)(x - x^3) = F\cos(\Omega t), \tag{A.1}$$

that is, the double-well potential used extensively throughout this book. The circuit used here was designed by translating the block diagram into basic electric circuit components consisting of resistors, capacitors, operational amplifiers, and multiplier chips (Carlson and Gisser, 1981; Matsumoto, Chua, and Tanaka, 1984).

In the circuit, input from a function generator is modulated using a voltage controlled amplifier constructed from a multiplier chip as shown on the far left of the circuit in Figure A.1. This is analogous to direct, rather than transmissible, excitation. The forcing input is then added to the position, velocity, and the cubic term through the use of a summing amplifier. Further details of this modeling can be found in Ref. (Trickey, 1997). A key section of the circuit consists of the two multiplier chips at the far right of Figure A.1, which constitute the cubic term in Duffing's equation A.1. The circuit was tuned to have a linear natural frequency of about 11.7 Hz (i.e., a time scale about an order of magnitude faster than the mechanical systems discussed in this book) and stable equilibrium solutions at about ± 1 V. We again use $\Omega = \omega/\omega_n$ as the control parameter, and a similar process of nondimensionalization to the mechanical oscillator was conducted. A slight asymmetry in the two natural frequencies (and hence potential energy) is unavoidable based on nominal resistor values.

The acquisition of experimental data was also accomplished using virtual instrument's (VIs) within the LabVIEW environment, with very little alteration from the LabVIEW set up to acquire data from the mechanical systems discussed earlier (e.g., Figure 6.4).

A.2 Typical Responses

For relatively small forcing conditions, the circuit will exhibit the standard linear features first outlined in Chapter 2. The focus here will be on features in which global behavior comes in to play. As a spot check, consider the chaotic attractor taken directly from the circuit output and shown in Figure A.2. This response is remarkably similar to the chaotic attractor obtained from the mechanical system (Figure 9.4). Clearly, similar folding and stretching processes are at work (Matsumoto, 1987).

Let's now consider what happens to transients as they evolve from initial conditions to persistent steady-state behavior. The region of main resonant hysteresis is chosen, and Figure A.3 shows a typical amplitude response diagram. The saddle-node bifurcation of interest occurs around $\Omega = 0.885$, where four period-one attractors become two, and these results pertain to a smaller forcing amplitude ($F = 0.15$) than that used in Figure A.2. Within each potential energy well, we can consider these coexisting attractors to be the resonant and nonresonant branches.

The method of stochastic interrogation (Cusumano and Kimble, 1995) was used to generate initial conditions, which were then assigned a basin when the trajectory converged to within a tolerance of one of the predetermined steady-state solutions. Figure A.4 shows a typical experimental result with the four shades of gray corresponding to (a) the

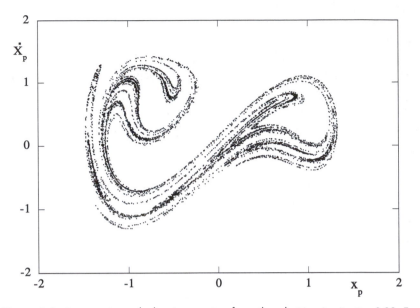

Figure A.2: An experimental chaotic attractor from the electric circuit. $\Omega = 0.82$, $F = 0.65$, and ζ (measured) $= 0.019$.

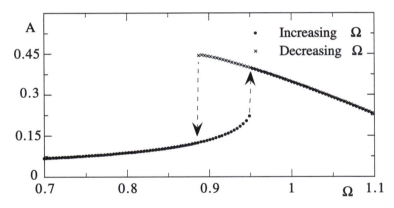

Figure A.3: A amplitude response diagram showing the region of main resonant hysteresis measured from the experimental circuit in terms of the amplitude of response $A = (X_{max} - X_{min})/2$. The forcing amplitude is $F = 0.15$.

four solutions when $\Omega = 0.91$ and (b) two coexisting solutions when $\Omega = 0.88$.

This type of behavior is also present in numerical simulations. Figure A.5 shows a bifurcation diagram (qualitatively similar to the

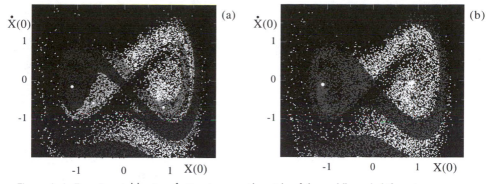

Figure A.4: Experimental basins of attraction on either side of the saddle-node bifurcation. (a) $\Omega = 0.91$; (b) $\Omega = 0.88$.

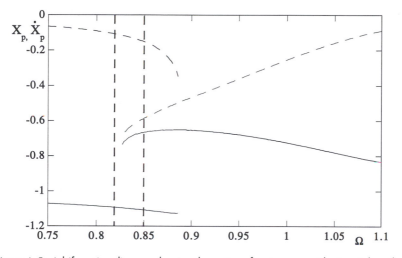

Figure A.5: A bifurcation diagram showing the region of main resonant hysteresis based on numerical simulation. X_p: continuous; $\dot{X}_p$: dashed.

amplitude response diagram of Figure A.3) again highlighting the region of main resonant hysteresis. We note that the equation of motion underlying these results was actually devised to model a slightly (quantitatively) different electric circuit than the one just described. Because of the difference, primarily in the damping coefficents, we see a shift in the frequencies where the saddle-nodes, and hence, hysteresis occurs, and there is also a difference in the scaling used for velocity (due to the use of arbitrary constants in the operational amplifiers). Figure A.6 shows the same qualitative results for the basins of attraction based on

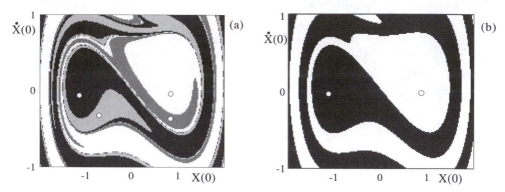

Figure A.6: Numerical basins of attraction on either side of the saddle-node bifurcation. (a) $\Omega = 0.85$; (b) $\Omega = 0.82$.

Equation A.1. Note that although the *measurement* of damping is subject to the usual experimental effects such as noise, because the circuit is based on simple integrator elements, we can be relatively confident (in contrast to typical mechanical systems) in the modeling of damping as a linear viscous element. We also note that it is simple to extract state variables in this application and, hence, recourse to time-lag embedding is unnecessary.

The values of nondimensional frequency ratio at which the numerical basins were computed are also indicated in Figure A.5 by the vertical dashed lines. These results are based on integrating from a regular grid of initial conditions, and the similarity to the (transmissibly forced) results of Figure 14.1 is clear. This figure also helps to clarify (visually) the shades of gray in the experimental results.

Regardless of the specific form of the equation of motion or the parameter values used, it is clear that the same local and global dynamic processes are at work in both the experimental and numerical data.

A.3 Bottlenecking

Figure A.7 contains the transient length information shown for a frequency just after the saddle-node bifurcation (i.e., corresponding to Figure A.4(b) and A.6(b)), where the shades of gray now represent the

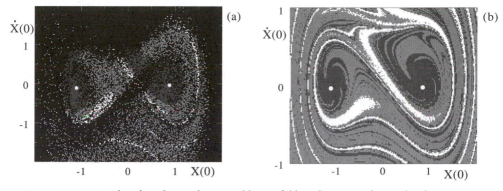

Figure A.7: Transient lengths indicating how unstable manifolds and remnant solutions slow the progress of transients toward the stable solutions. (a) Experimental results based on stochastic interrogation; parameters same as Figure A.4(b). (b) Numerical results based on a regular grid of starts; parameters same as Figure A.6(b).

following rate of attraction grouped in terms of the number of forcing cycles prior to convergence:

- fast (dark gray) – experimental (0–5), simulation (0–9);
- medium (light gray) – experimental (6–14), simulation (10–19);
- slow (white) – experimental (15+), simulation (20+).

Again, these results relate to slightly different cases, but the tendencies are the same. Not surprisingly, we see a general trend of points starting close to a fixed point tending to return there fairly quickly, although the shading scheme does allow for some of these (shaded dark) initial conditions to be rapidly attracted to a remote fixed point. We note that those initial conditions that take a relatively long time to settle appear to fall roughly along the separatrix in the corresponding basin diagrams. This region not only contains the unstable fixed points (between the resonant and nonresonant branches within each well and a small hilltop solution) but also their stable manifolds. Some of these slow, or long-lived, transients are caused by an interesting interaction of local and global events.

Some of the white data points in Figure A.7 are slowed by their encounter with the ghost, or remnant, of the previously existing attractor. They are not unstable orbits. The trajectories slow down as they

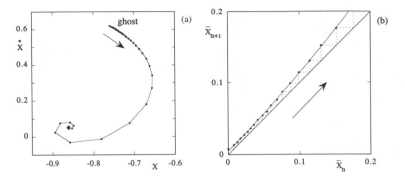

Figure A.8: (a) A typical experimental Poincaré section highlighting the slowing down of the trajectory close to a ghost solution; (b) the corresponding cobweb diagram based on a one-dimensional manifold projection. $\bar{X}$ is measured from an estimate of the ghost location.

pass by the remnant of the former stable fixed point on an essentially one-dimensional (1-D) manifold. This phenomenon of *bottlenecking* is also present in the intermittent form of chaos, although there the trajectory is repeatedly reinjected through the bottlenecking region (Ott, 1993; Kim et al., 1997). It is stochastic interrogation that allows this behavior to be observed experimentally in this system; an isolated ghost solution can only be accessed by a small subset of initial conditions. The dwelling of the trajectory in the vicinity of the former fixed point (now destroyed at a saddle-node) can be seen in the Poincaré section shown in Figure A.8(a) before it slides off and toward the stable fixed point. This behavior is occasionally observed during random starting conditions on the oscilloscope in the laboratory.

Figure A.8(b) shows how the underlying 1-D map does not cross the 45^0 line. An intersection would imply $x_n = x_{n+1}$, that is, a fixed point, whose stability is determined by the slope at intersection – a geometric interpretation of a characteristic multiplier (CM). We see the long sequence of iterations during the traverse through the bottleneck.

It is also interesting to note that the degree of slowing down due to bottlenecking follows an inverse square root scaling with the distance from the ghost solution. This theory, based on generic saddle-node bifurcations in Refs. (Strogatz, 1994; Wiggins, 1990) and in the context of intermittency in Refs. (Moon, 1992; Ott, 1993; Bergé et al., 1984), states that the time required to pass through a bottleneck region (T) induced by a saddle-node remnant scales with the inverse root of the change in

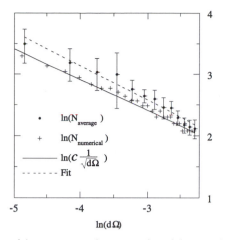

Figure A.9: A log plot of the experimental, numerical, and theoretical results. The slopes of −0.52 (experimental) and −0.5 (theoretical) are in close agreement. The error bars on the experimental data represent one standard deviation.

bifurcation parameter (p):

$$T \propto \frac{1}{\sqrt{p_c - p}}. \tag{A.2}$$

A nondimensional summary of this scaling law, where $p_c - p \equiv d\Omega$, is shown in Figure A.9, together with some numerical results (based on the numerical integration of Equation A.1), included for comparative purposes (Trickey and Virgin, 1998). The number N relates to the speed of transit. This is just a specific example of nonlinear behavior with a global aspect. Similar bottlenecking phenomena is also displayed by the mechanical analog.

This circuit can also be used to characterize unstable periodic orbits, nonstationary effects and the control of chaos, and, in fact, the whole of the spectrum of nonlinear behavior encountered throughout this book.

Appendix B

A Continuous System

B.1 Introduction

All of the preceding chapters have dealt with single-degree-of-freedom oscillators, generally with a periodic excitation and, hence, a three-dimensional phase space. Many real dynamical systems are continuous, modeled by partial differential equations with both space and time as independent variables. Despite the fact that the dynamics often take place on a relatively low-order subspace of the (infinite) phase space of the full system, there are still many situations in which an analysis in a high-order space is necessary. For continuous systems such as beams and plates (see Chapter 4), modal analysis has proved to be a powerful technique for extracting the dominant dynamic characteristics from complex systems, especially for linear systems. In a theoretical context, Galerkin's method can be used to reduce a partial differential equation into a set of coupled ordinary differential equations, which can then be analyzed using standard techniques. The success with which a reduced-order model captures the full range of behavior is a very complicated issue (especially in fluid mechanics (Lorenz, 1963; Ruelle and Takens, 1971)), but, for example, a continuous beam excited close to its fundamental natural frequency will display behavior dominated by the lowest mode, and hence a lumped parameter model will likely be good enough in an engineering context.

Some of the earliest studies in chaos were generated by the consideration of thin beams, which under certain circumstances could be very successfully modeled by Duffing's equation (see (Moon, 1992) and Chapter 4). The presence of multiple equilibria and periodic excitation provided conditions under which a wide range of nonlinear behavior could be observed and measured. In this appendix, we take a brief look at a continuous (i.e., high-order) experimental system that displays behavior that is qualitatively similar to the single-degree-of-freedom examples encountered earlier. The practical context for this example occurs in certain aerospace systems where thin metal panels are subject to intense acoustic excitation and are often in a postbuckled equilibrium configuration owing to thermal effects (Tauchert, 1991).

The theoretical treatment is rather involved, and the reader is referred to Refs. (Murphy, Virgin, and Rizzi, 1996a; Murphy, Virgin, and Rizzi, 1997; Murphy, 1994) for more details. Here, we shall concentrate on experimental results and try to contrast the similarities and differences with some of the (low-order) results presented earlier in this book.

B.2 Underlying Statics

In the Duffing track-cart system, the stiffness was effectively fixed by the shape of the track. Now, with this continuous system, the stiffness (and hence potential energy function) can be conveniently adjusted by the temperature loading with the twin-well configuration supplying the main emphasis.

A clamped plate has an infinite number of degrees of freedom and, hence, natural frequencies and mode shapes. However, a typical response (in bending) for a plate with an aspect ratio close to unity will tend to be dominated by the lower modes (e.g., the (1,1) mode consisting of a cosine wave in each direction). Under the influence of an applied thermal loading condition, the plate will experience compression caused by its constrained edges (Johns, 1965). This will tend to reduce the effective lateral stiffness and, therefore, natural frequencies. For an initially flat plate, a critical temperature will cause buckling, and the plate will bifurcate into either of the (symmetric) postbuckled equilibrium configurations available. However, structural systems of this kind will always have some kind of imperfection (often an initial geometric deflection)

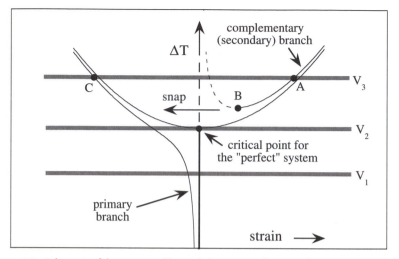

Figure B.1: Schematic of the static equilibrium behavior as a function of temperature rise. The dashed curve starting from point B is the unstable equilibrium path.

such that the symmetry is broken and the plate continuously deflects with an increase in the rate of deflection in the vicinity of the critical temperature for the underlying *perfect* geometry. The plate thus has a preferred direction of buckling displacement. The presence of a complementary postbuckled equilibrium means that if sufficient force is applied, it is possible to deflect the plate into this other direction, although this would not occur in a *natural* loading history.

Figure B.1 provides a conceptual summary of the underlying statics, and indeed, we can immediately identify the potential level V_3 as providing the analogy to the earlier double-well systems. Since we are primarily interested in experimental work, we shall focus attention on the imperfect system. In this case, instead of the bifurcation, the only distinct instability is a saddle-node at temperature point B where the secondary solution comes into existence. Thus, a decrease in temperature from high values (starting at point A) passing through point B will result in a dynamic snap as the system inevitably jumps to the primary equilibrium path (rather like the way a cookie sheet can make this sound as it cools in the oven) (Virgin, 1986).

Here we shall focus on the response of the plate to narrow-band (effectively harmonic) acoustic excitation (Beranek, 1954; Kinsler et al., 1982) at temperatures above the critical value (e.g., V_3 in Figure B.1).

At relatively low temperatures (e.g., V_1) the response is not significantly different from that of a linear system (provided the lateral forcing does not induce significant membrane effects). At postcritical temperatures, the potential energy of the system can be approximated by the double-well configuration (see Figure 5.3) with the (slightly asymmetric) minima representing the two postbuckled equilibrium configurations. Hence, subjecting the system to periodic lateral excitation, we might not be surprised to observe nearly linear responses for small levels of excitation (but about a nonlinear equilibrium position) with the possibility of *escape* and persistent snap-through behavior for higher levels of excitation. This behavior is similar to that of the cart-track system, despite the high-order nature of the continuous panel.

B.3 Experimental Description

The experiment to be described was conducted at the Thermal Acoustic Fatigue Apparatus (TAFA) at NASA Langley Research Center, a facility capable of providing combined thermal and acoustic loads. A full description of this particular experimental setup, including boundary conditions (and their evaluation), can be found in Refs. (Murphy, Virgin, and Rizzi, 1996a; Murphy, Virgin, and Rizzi, 1997), and the facility is described in Ref. (Clevenson and Daniels, 1992).

Initially consider a thin panel with dimensions 0.381 m × 0.305 m × 3.175 mm and made of AISI stainless steel as shown in Figure B.2. High-temperature strain gauges were used (near the lower left corner of the plate) together with thermocouples, and sophisticated data acquisition and analysis software were utilized.

B.3.1 Free Response

For nondimensionalization purposes, we will use the characteristics of the plate at ambient (room) temperature conditions, assuming a perfectly flat geometry but allowing for a finite in-plane edge stiffness. It was found that this factor played an important role in the subsequent correlation between theory and experiments.

Under fully clamped boundary conditions, but allowing for a little (experimentally determined) in-plane slippage, the lowest natural

Figure B.2: Photograph of the panel.

frequency of the panel $(\omega_n|_{\Delta T=0})$ at ambient temperature conditions ($\approx 20°C$) was measured at about 220 Hz (242 Hz in simulations), and the static buckling temperature (ΔT_{cr}^f) was estimated at around 86°C above ambient.

Figure B.3 shows a plot of the relationship between the temperature and lowest natural frequency. A laser velocity vibrometer used to extract mode shapes indicated that this frequency corresponded to a mode shape dominated by a single cosine mode in each direction. The axes have been nondimensionalized such that the temperature is relative to the critical (buckling) temperature, and the natural frequency is relative to the baseline natural frequency at the ambient temperature, that is,

$$\bar{\omega}_n = \frac{\omega_n}{\omega_n|_{\Delta T=0}}, \qquad (B.1)$$

using the values given above.

The underlying theory for the perfectly flat geometry is based on a classical eigenvalue analysis, and plotting the frequency squared is appropriate given the relationship between natural frequency and lateral stiffness. The postbuckled increase in frequency has been observed in a number of related structural systems (Lurie, 1952; Virgin, 1985).

When the geometric imperfection (measured experimentally using the Southwell Plot (Croll and Walker, 1972)) is included in the

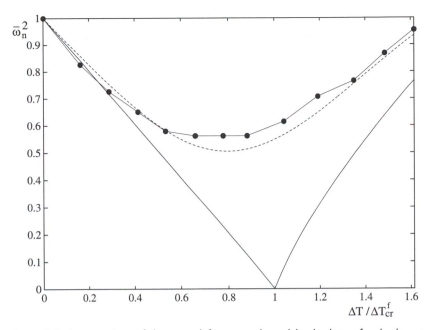

Figure B.3: A comparison of the natural frequency–thermal load relation for the lowest frequency, i.e., the (1,1) mode, for the clamped panel. • – experimental data; continuous line – theoretical results based on an initially perfectly flat panel; dashed line – theoretical results based on an initially imperfect geometry.

analysis, the eigenproblem must be solved numerically. The dashed line in Figure B.3 shows the result, which agrees well with the experimental results, and hence, in practice, we will not observe a distinct buckling phenomena with a zero natural frequency. Further modeling challenges are presented by the small amount of in-plane flexibility in the clamped boundaries and the issues surrounding the uniformity of thermal loading through the panel and its support frame, as discussed in Ref. (Murphy, 1994). The shape of the underlying potential energy well thus changes with temperature, and, although not shown in Figure B.3, another natural frequency appears owing to the complementary (imperfect) equilibrium path (A–B in Figure B.1) for high temperatures.

Again, the modeling of damping is a complex issue, especially considering the finite in-plane edge stiffness at the boundaries (to be discussed in more detail a bit later). However, Figure B.4 shows a typical transfer function (i.e., input/output as a function of frequency) obtained

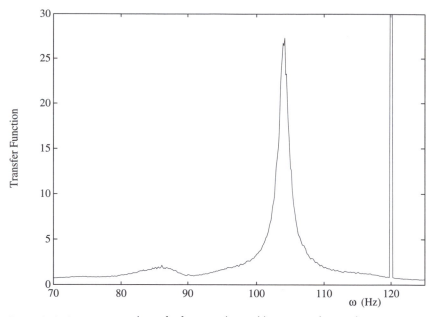

Figure B.4: An experimental transfer function obtained by exciting the panel at 86 Hz at a temperature of $\Delta T = 11°C$.

from the experimental panel. The half-power method (see Section 2.5.2) was used to extract a damping ratio of about 1% ($\zeta = 0.01$), and this was seen to be broadly independent of temperature. We know that this is not quite true since the definition of damping ratio (Equation 2.13) has the natural frequency (which does depend on temperature) built into it. This value is the damping ratio for the first (dominant) mode; in the corresponding theoretical study (Murphy, Virgin, and Rizzi, 1996a) the higher mode damping ratios were distributed and scaled according to their resonant frequencies. The spike at 120 Hz in Figure B.4 is the second harmonic of 60 Hz line noise, which was not even an issue for the characteristic frequencies of the Duffing cart-track system.

B.4 Typical Forced Responses

Before displaying the response to narrow-band excitation, we note that the following results are based on an experimental panel with half the thickness (i.e., 1.58 mm) of the one used in the previous section. This was chosen to promote flexibility and (ultimately) cross-well behavior.

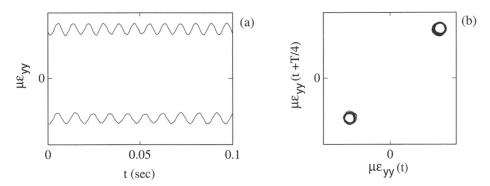

Figure B.5: Small-amplitude periodic behavior about both postbuckled positions using micro-strain and time-lag embedding. (a) Time series and (b) phase projections.

Given this change in geometry, the baseline parameters of the (less stiff) panel are now $\omega_n|_{\Delta T=0} = 122$ Hz and $\Delta T_{cr} \approx 21.5°$C. Thus the natural frequency scales linearly with the thickness and the critical temperature scales with the thickness squared – the slight discrepancy was due to fitting the finite edge stiffness from the experiments. Armed with this information about the underlying characteristics of the system, we first consider rather small levels of excitation, displaying the response in terms of strain. Figure B.5 shows the two competing steady-state oscillations superimposed (i.e., motion contained within each potential energy well as both time series and phase projections) corresponding to small amplitude motion about A and C in Figure B.1. These results were recorded at a temperature of $\Delta T / \Delta T_{cr}^f = 1.76$ with an excitation level of 130 dB at 120 Hz (quite near the lowest natural frequency at this temperature). The second attractor (which has a somewhat smaller basin of attraction) is attained by giving a variety of perturbations to access different areas of the initial condition space. The mild asymmetry (see points A and C in Figure B.1) causes a slight difference in the positive and negative strains as well as a mild difference in the two periods of response. The reconstructed phase projection is based on the time-lag embedding the strain data using a quarter-cycle delay. This kind of small-amplitude, periodic behavior possesses a power spectrum with a dominant spike at the forcing frequency (actually based on the positive strain oscillation) and a periodic autocorrelation function (see Figure B.6).

On further increase in excitation, we observe behavior typified by Figure B.7. Here, the response was recorded at a temperature of

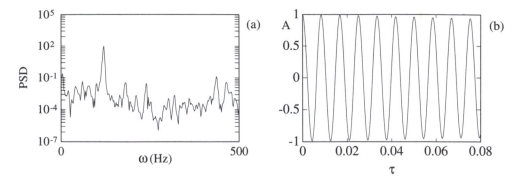

Figure B.6: (a) Power spectrum and (b) autocorrelation function based on the response shown in Figure B.5.

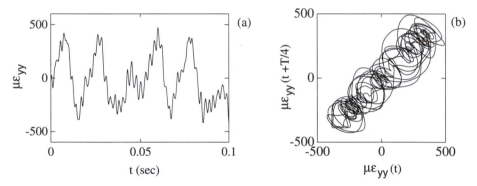

Figure B.7: (a) Time series and (b) phase projection (using time-delayed strain) for the aperiodic response of the panel.

$\Delta T/\Delta T_{cr}^{f} = 1.53$ at an excitation level of 155 dB at 115 Hz (with a lowest natural frequency of 106 Hz at this temperature). Note that although the temperature has been reduced in going from Figure B.5 to Figure B.7, since we are heavily in the postcritical regime, the effective natural frequency is somewhat counterintuitively reduced. This erratic oscillation certainly has the appearance of chaos, although the time series and phase projection are a subset of the full behavior. Again, this plot shows the measured strain as a function of time. Observation of the experimental response indicated that this motion did indeed correspond to snap-through behavior as the oscillation intermittently moved about and between the buckled equilibrium configurations.

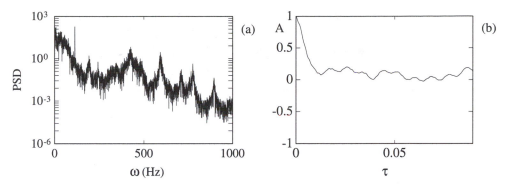

Figure B.8: (a) Power spectrum and (b) autocorrelation function based on the response shown in Figure B.7.

The power spectrum and autocorrelation (Figure B.8) give further support to labeling this behavior as chaotic. However, the other measures, including dimension and Lyapunov exponents, are more difficult to apply since they require potentially enormous amounts of data to establish statistical properties for high-order systems. The use of surrogate data can be useful here to help distinguish between deterministic and stochastic behavior. It can be mentioned, however, that Lyapunov exponent computations for both simulations and experimental data for this apparently aperiodic behavior produced two positive values, although repeated difficulty was encountered in ensuring convergence in the dimension and embedding calculations (Murphy, Virgin, and Rizzi, 1996a).

A full complement of numerical results were generated based on a nine-cosine-mode (three in each direction) model, involving Hamilton's principle, Galerkin's method, Airy stress functions, a Southwell plot, etc. Most of the results of the experiment were verified, although the effects of noise were apparent in the experimental results (Murphy, Virgin, and Rizzi, 1996a).

B.5 Snap-Through Behavior

Finally, the snap-through behavior caused by the transition from small-amplitude motion to cross-well large-amplitude motion is also very similar to the escape study described in Chapter 10.

Figure B.9 shows two examples of nonstationary snap-through behavior where the sound pressure level (SPL) is gradually swept up and then down to produce a nonstationary transition through cross-well behavior. The results are based on $\Delta T / \Delta T_{cr}^f = 1.76$ with a *baseline* forcing of 130 dB at 120 Hz, (i.e., as in Figure B.5). In both parts, the SPL is ramped from 130 dB up to 150 dB and then back down to 130 dB, although not necessarily at the same rate because the sweep was controlled manually. In part (a), the motion begins as a small-amplitude periodic oscillation until experiencing a burst of snap-through behavior followed by motion about the other equilibrium configuration (the adjacent potential energy well). Note the *beating* effect prior to the cross-well motion. This has been observed in generic systems as a nonstationary precursor to the jump phenomenon (Thompson and Virgin, 1986) (see also Figure 8.9). In part (b) the system is initiated with nominally the same conditions. After the burst of cross-well motion, the system, upon subsequent reduction in the applied SPL, settles back to small-amplitude motion about the original equilibrium. We can also note that the post–cross-well behavior in part (a) appears to be a somewhat larger motion than the original motion in the other well. Despite the symmetric nature of the underlying potential energy curve it is quite likely that a resonant motion has been picked up, a motion perhaps coexisting with a small-amplitude, nonresonant branch in much the same way as the multiple solution regime in Figure 8.5.

In a practical context, it might be very useful to have an idea of the regions of parameter space that typically lead to snap-through behavior, akin to the escape characteristics in Figure 10.8. Fixing the (postcritical) temperature at $\Delta T / \Delta T_{cr}$ thus determines the fundamental natural frequency ω_n (snap-through behavior in the vicinity of primary resonance is of most interest). For a given frequency of excitation, the SPL is then quasi-statically increased until the first occurrence of snap-through behavior occurs, that is, the system traverses the potential energy hilltop associated with the unstable (almost flat) equilibrium configuration. The frequency is then incremented (on an effectively slower evolution rate than the one used to produce Figure B.9), and the procedure is repeated. Figure B.10 shows a summary of 30 such runs plotted against nondimensional frequency, clearly separating regions of parameter space into *snap-through* and *no snap-through* regions (Murphy, Virgin, and Rizzi, 1996b).

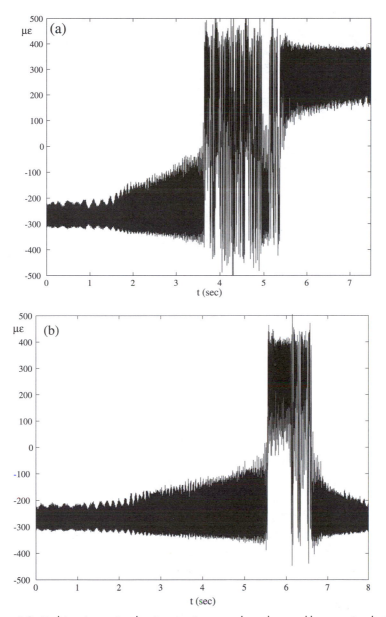

Figure B.9: Evolving time series showing transient snap-through caused by sweeping the SPL from 130 dB → 150 dB → 130 dB. (a) The motion was initiated around the secondary equilibrium but then settled around the primary equilibrium. (b) The motion was initiated around the secondary equilibrium and returned there after a burst of transient snap-through.

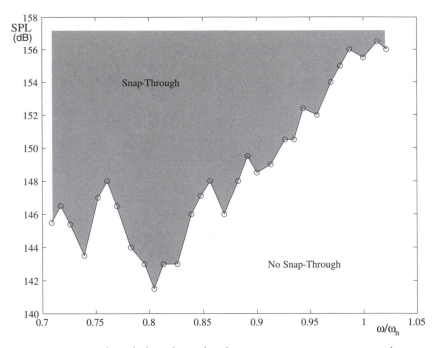

Figure B.10: Snap-through boundary plotted in parameter space using tonal inputs $\Delta T/\Delta T_{cr} = 1.95$, $\omega_n \approx 111$ Hz at this temperature.

This figure bears some resemblance to the lower boundary of Figure 10.8, which was also based on a quasi-static ramping of the forcing conditions. Of course, there are a host of more complicated issues here, but nonetheless both systems are most likely to experience escape or snap-through when strongly forced close to resonance, which is entirely to be expected owing to the amplitude magnification effects. Also, in both cases, the minimum force level corresponds to a frequency just below the fundamental natural frequency, which again is anticipated from the softening spring effect of the nonlinearity.

The practical motivation for avoiding large-amplitude snap-through behavior in terms of sonic fatigue mitigation is clear (Ng and White, 1988; Clarkson, 1994).

Bibliography

Abarbanel, H. D. I., Brown, R., Sidorowich, J., and Tsimring, L. (1993). The analysis of observed chaotic data in physical systems. *Reviews of Modern Physics*, 65:1331–1392.

Abraham, R. H. and Shaw, C. D. (1982). *Dynamics: The Geometry of Behavior*. Aerial Press.

Arnold, T. W. and Case, W. (1982). Nonlinear effects in a simple mechanical system. *American Journal of Physics*, 50:220–224.

Arnold, V. I. (1965). Small denominators, I: Mappings of the circumference into itself. *AMS Translation Series 2*, 46:213–284.

Arnold, V. I. (1988). *Geometrical Methods in the Theory of Ordinary Differential Equations*. Springer-Verlag.

Arrechi, F. T., Badii, R., and Politi, A. (1984). Scaling of first passage times for noise induced crises. *Physics Letters A*, 103:3–7.

Badii, R. and Politi, A. (1985). Statistical description of chaotic attractors: The fractal dimension. *Journal of Statistical Physics*, 40:725–750.

Baker, G. L. and Gollub, J. P. (1996). *Chaotic Dynamics: An Introduction*. Cambridge University Press, 2nd ed.

Banbrook, M., Ushaw, G., and McLaughlin, S. (1997). How to extract Lyapunov exponents from short and noisy time series. *IEEE Transactions on Signal Processing*, 45:1378–1382.

Bayly, P. V. and Virgin, L. N. (1993a). An empirical study of the stability of periodic motion in the forced spring–pendulum. *Proceedings of the Royal Society of London A*, 443:391–408.

Bayly, P. V. and Virgin, L. N. (1993b). An experimental study of an impacting pendulum. *Journal of Sound and Vibration*, 162:1–11.

Bayly, P. V. and Virgin, L. N. (1994). Practical considerations in the control of chaos. *Physical Review E*, 50:604–607.

Bayly, P. V., Virgin, L. N., Gottwald, J. A., and Dowell, E. H. (1994). Stability measurements in nonlinear mechanical experiments guided by dynamical systems theory. In *Proceedings of the IUTAM Symposium: Nonlinearity and Chaos in Engineering Dynamics*. Wiley.

Bazant, Z. P. and Cedolin, L. (1991). *Stability of Structures*. Oxford University Press.

Beckwith, T. G., Marangoni, R. D., and Lienhard, J. H. (1993). *Mechanical Measurements*. Addison-Wesley.

Begley, C. J. and Virgin, L. N. (1997). A detailed study of low-frequency periodic behavior of a dry friction oscillator. *Journal of Dynamic Systems, Measurement and Control*, 119:491–497.

Begley, C. J. and Virgin, L. N. (1998). Impact response and the influence of friction. *Journal of Sound and Vibration*, 211:801–818.

Bendat, J. S. and Piersol, A. G. (1986). *Random Data: Analysis and Measurement Procedures*. Wiley.

Beranek, L. L. (1954). *Acoustics*. McGraw-Hill.

Bergé, P., Pomeau, Y., and Vidal, C. (1984). *Order within Chaos: Towards a Deterministic Approach to Turbulence*. Wiley, New York.

Birkhoff, G. (1941). Some unsolved problems of theoretical dynamics. *Science*, 94:598–600.

Bogoliubov, N. N. and Mitropolsky, Y. A. (1961). *Asymptotic Methods in the Theory of Nonlinear Oscillations*. Gordon and Breach.

Brandstater, A. and Swinney, H. L. (1987). Strange attractors in weakly turbulent Couette–Taylor flow. *Physical Review A*, 35:2207–2220.

Broomhead, D. S. and King, G. P. (1986). Extracting qualitative dynamics from experimental data. *Physica D*, 20:217–236.

Brown, R., Bryant, P., and Abarbanel, H. D. I. (1990). Computing the Lyapunov spectrum of a dynamical system from an observed times series. *Physical Review A*, 43:2787–2806.

Brunsden, V., Cortell, J., and Holmes, P. J. (1989). Power spectra of chaotic vibrations of a buckled beam. *Journal of Sound and Vibration*, 130:1–25.

Burton, T. D. (1994). *Introduction to Dynamic Systems Analysis*. McGraw-Hill.

Carlson, A. B. and Gisser, D. G. (1981). *Electrical Engineering: Concepts and Applications*. Addison-Wesley.

Carr, J. (1981). *Applications of Centre Manifold Theory*. Springer-Verlag.

Casdagli, M., Eubank, S., Framer, J. D., and Gibson, J. (1991). State space reconstruction in the presence of noise. *Physica D*, 51:52–98.

Childs, D. (1993). *Turbomachinery Rotordynamics*. Wiley Interscience.

Chin, W., Ott, E., Nusse, H. E., and Grebogi, C. (1994). Grazing bifurcations in impact oscillators. *Physical Review E*, 50:4427–4444.

Chugani, M., Samant, A., and Cerna, M. (1998). *LabVIEW Signal Processing*. Prentice-Hall.

Clarkson, B. L. (1994). Review of sonic fatigue technology. Technical report, NASA Contract Report 4587.

Clevenson, S. A. and Daniels, E. F. (1992). Capabilities of the thermal acoustic fatigue apparatus. Technical report, NASA Technical Memorandum 104106.

Cochin, I. and Plass, H. P. (1990). *Analysis and Design of Dynamic Systems*. Harper and Row.

Collinge, I. R. and Ockendon, J. R. (1979). Transition through resonance of a Duffing oscillator. *SIAM Journal of Applied Mathematics*, 37:350–357.

Conner, M. D., Tang, D. M., Dowell, E. H., and Virgin, L. N. (1997). Nonlinear behavior of a typical airfoil section with control surface freeplay: A numerical and experimental study. *Journal of Fluids and Structures*, 11:89–109.

Cooley, T. W. and Tukey, J. W. (1965). An algorithm for the machine calculation of complex Fourier series. *Mathematical Communications*, 19:297–301.

Croll, J. G. A. and Walker, A. C. (1972). *Elements of Structural Stability*. Wiley.

Cusumano, J. P. (1990). *Low-dimensional, Chaotic, Nonplanar Motions of the Elastica: Experiment and Theory*. PhD thesis, Cornell University.

Cusumano, J. P. and Kimble, B. W. (1994). Experimental observation of basins of attraction and homoclinic bifurcation in a magneto-mechanical oscillator. In *Proceedings of the IUTAM Symposium: Nonlinearity and Chaos in Engineering Dynamics*. Wiley.

Cusumano, J. P. and Kimble, B. W. (1995). A stochastic interrogation method for experimental measurements of global dynamics and basin evolution: Application to a two-well oscillator. *Nonlinear Dynamics*, 8:213–235.

den Hartog, J. P. (1930). Forced vibrations with combined Coulomb and viscous damping. *Transactions of the ASME*, 53:107–115.

den Hartog, J. P. (1984). *Mechanical Vibrations*. Dover.

Ding, M., Grebogi, C., Ott, E., Sauer, T., and Yorke, J. A. (1993). Plateau onset for correlation dimension: When does it occur? *Physical Review Letters*, 70:3872–3875.

Ditto, W. L., Rauseo, S. N., and Spano, J. L. (1990a). Experimental control of chaos. *Physical Review Letters*, 65:3211–3214.

Ditto, W. L., Spano, M. L., Savage, H. T., Rauseo, S. N., Heagy, J., and Ott, E. (1990b). Experimental observation of a strange nonchaotic attractor. *Physical Review Letters*, 65:533–536.

Doeblin, E. O. (1990). *Measurement Systems: Applications and Design*. McGraw-Hill.

Doedel, E. J. (1986). *AUTO – Software for continuation and bifurcation problems in ordinary differential equations*. California Institute of Technology.

Dowell, E. H. (1975). *Aeroelasticity of Plates and Shells*. Noordhoff.

Dowell, E. H. and Pezeshki, C. (1986). On the understanding of chaos in Duffing's equation including a comparison with experiment. *Journal of Applied Mechanics*, 53:5–9.

Dressler, U. and Nitsche, G. (1992). Controlling chaos using time delay coordinates. *Physical Review Letters*, 68:1–4.

239

Duffing, G. (1918). *Erzwungene Schwingungen bei veranderlicher Eigenfrequenz.* F. Vieweg u. Sohn.

Ebert, T. F. (1984). *Estimation and Control of Systems.* Van Nostrand.

Eckmann, J.-P. and Ruelle, D. (1992). Fundamental limitations for estimating dimensions and Lyapunov exponents in a dynamical system. *Physica D*, 56:185–187.

Ehrich, F. F. (1991). Some observations of chaotic vibration phenomena in high-speed rotor dynamics. *Journal of Vibration and Acoustics*, 113:50–57.

Eschenazi, E., Solari, H. G., and Gilmore, R. (1989). Basins of attraction in driven dynamical systems. *Physical Review A*, 39:2609–2627.

Ewins, D. J. (1984). *Modal Testing: Theory and Practice.* Research Studies Press.

Farmer, J. D., Ott, E., and Yorke, J. A. (1983). The dimensions of chaotic attractors. *Physica D*, 7:153–180.

Feeney, B. F. and Liang, J. W. (1996). A decrement method for the simultaneous estimation of Coulomb and viscous friction. *Journal of Sound and Vibration*, 195:149–154.

Feigenbaum, M. J. (1978). Quantitative universality for a class of nonlinear transformations. *Journal of Statistical Physics*, 19:25–32.

Feynman, R. P., Leighton, R. B., and Sands, M. (1963). *The Feynman Lectures on Physics.* Addison-Wesley.

Forbes, L. K. (1989). A series analysis of forced transverse oscillations in a spring-mass system. *SIAM Journal of Applied Mathematics*, 49:704–719.

Frazer, A. M. and Swinney, H. L. (1986). Independent coordinates for strange attractors from mutual information. *Physical Review A*, 33:1134–1140.

Gear, C. W. (1971). *Numerical Initial Value Problems in Ordinary Differential Equations.* Prentice-Hall.

Gleick, J. (1987). *Chaos.* Penguin.

Goldstein, H. (1980). *Classical Mechanics.* Addison-Wesley.

Golubitsky, M. and Shaeffer, D. G. (1985). *Singularities and Groups in Bifurcation Theory.* Springer-Verlag.

Goodwin, G. C. and Payne, R. L. (1977). *Dynamics System Identification: Experiment Design and Data Analysis.* Academic Press.

Gottlieb, H. P. W. (1997). Exact mimicry of nonlinear oscillatory potential motion: Nonuniqueness of isodynamical tracks. *Journal of Sound and Vibration*, 204:519–532.

Gottwald, J. A., Virgin, L. N., and Dowell, E. H. (1992). Experimental mimicry of Duffing's equation. *Journal of Sound and Vibration*, 148:447–467.

Gottwald, J. A., Virgin, L. N., and Dowell, E. H. (1995). Routes to escape from an energy well. *Journal of Sound and Vibration*, 187:133–144.

Grassberger, P. and Procaccia, I. (1983). Measuring the strangeness of strange attractors. *Physica D*, 9:189–208.

Grebogi, C., Ott, E., Pelikan, S., and Yorke, J. A. (1984). Strange attractors that are not chaotic. *Physica D*, 13:216–268.

Grebogi, C., Ott, E., and Yorke, J. A. (1983). Crises, sudden changes in chaotic attractors and transient chaos. *Physica D*, 7:181–200.

Grebogi, C., Ott, E., and Yorke, J. A. (1986a). Critical exponents of chaotic transients in nonlinear dynamical systems. *Physical Review Letters*, 57:1284–1287.

Grebogi, C., Ott, E., and Yorke, J. A. (1986b). Metamorphoses of basin boundaries in nonlinear dynamical systems. *Physical Review Letters*, 56:1011–1014.

Grebogi, C., Ott, E., and Yorke, J. A. (1987). Chaos, strange attractors, and fractal basin boundaries in nonlinear dynamics. *Science*, 238:632–638.

Guckenheimer, J. and Holmes, P. J. (1983). *Nonlinear Oscillations, Dynamical Systems, and Bifurcations of Vector Fields*. Springer-Verlag.

Guckenheimer, J. and Worfolk, P. (1993). Dynamical systems: Some computational problems. In *Bifurcations and Periodic Orbits of Vector Fields*. Kluwer.

Gwinn, E. G. and Westervelt, R. M. (1986). Fractal basin boundaries and intermittency in the driven damped pendulum. *Physical Review A*, 33:4143–4155.

Harris, F. J. (1978). On the use of windows for harmonic analysis with the discrete Fourier transform. *Proceedings of the IEEE*, 66:51–83.

Hartman, P. (1964). *Ordinary Differential Equations*. Wiley.

Hayashi, C. (1964). *Nonlinear Oscillations in Physical Systems*. Princeton University Press.

Hénon, M. (1982). On the numerical computation of Poincaré maps. *Physica D*, 5:412–414.

Hilborn, R. C. (1994). *Chaos and Nonlinear Dynamics*. Oxford University Press.

Hirsch, M. W. and Smale, S. (1974). *Differential Equations, Dynamical Systems, and Linear Algebra*. Academic Press.

Holmes, P. J. (1977). Bifurcations to divergence and flutter in flow-induced oscillations. *Journal of Sound and Vibration*, 53:471–503.

Holmes, P. J. (1979). A nonlinear oscillator with a strange attractor. *Philosophical Transactions of the Royal Society of London*, 292:419–438.

Holmes, P. J. (1980). Introductory comments. In *New Approaches to Nonlinear Problems in Dynamics*. SIAM.

Holmes, P. J. (1990). Nonlinear dynamics, chaos, and mechanics. *Applied Mechanics Reviews*, 43:23–39.

Holmes, P. J. and Moon, F. C. (1983). Strange attractors and chaos in nonlinear mechanics. *Journal of Applied Mechanics*, 50:1021–1032.

Holmes, P. J. and Rand, D. A. (1976). The bifurcations of Duffing's equation: An application of catastrophe theory. *Journal of Sound and Vibration*, 44:237–253.

Hsu, C. S. (1987). *Cell-to-Cell Mapping: A Method of Global Analysis for Nonlinear Systems*. Springer-Verlag.

Huberman, B. A. and Crutchfield, J. P. (1979). Chaotic states of anharmonic systems in periodic fields. *Physical Review Letters*, 43:1743–1747.

Inman, D. J. (1994). *Engineering Vibration*. Prentice-Hall.

Jackson, E. A. (1989). *Perspectives of Nonlinear Dynamics*. Cambridge University Press.

Johns, D. J. (1965). *Thermal Stress Analysis*. Pergamon.

Jordan, D. W. and Smith, P. (1977). *Nonlinear Ordinary Differential Equations*. Clarendon Press.

Kaas-Petersen, C. (1987). Computation, continuation, and bifurcation of torus solutions for dissipative maps and ordinary differential equations. *Physica D*, 25:288–306.

Kao, Y. H., Huang, J. C., and Gou, Y. S. (1988). Routes to chaos in the Duffing oscillator with a single potential well. *Physics Letters A*, 131:91–97.

Kaplan, J. L. and Yorke, J. A. (1979). Chaotic behavior of multi-dimensional difference equations. In Peitgen, H. and Walther, H., editors, *Functional Differential Equations and the Approximations of Fixed Points*, volume 730. Springer-Verlag.

Karagiannis, K. and Pfeiffer, F. (1991). Theoretical and experimental investigations of gear-rattling. *Nonlinear Dynamics*, 2:367–387.

Katz, A. L. and Dowell, E. H. (1994). From single well chaos to cross well chaos: A detailed explanation in terms of manifold intersections. *International Journal of Bifurcation and Chaos*, 4:933–941.

Kelly, S. G. (1993). *Fundamentals of Mechanical Vibrations*. McGraw-Hill.

Kennel, M. B., Brown, R., and Abarbanel, H. D. I. (1992). Determining minimum embedding dimension using a geometrical construction. *Physical Review A*, 45:3403–3411.

Kesaraju, R. V. and Noah, S. T. (1994). Characterization and detection of parameter variations of nonlinear mechanical systems. *Nonlinear Dynamics*, 6:433–457.

Ketema, Y. (1992). A physical interpretation of Melnikov's method. *International Journal of Bifurcation and Chaos*, 2:1–9.

Kevorkian, J. and Cole, J. D. (1981). *Perturbation Methods in Applied Mechanics*. Springer-Verlag.

Kim, C. M., Yim, G. S., Kim, Y. S., Kim, J. M., and Lee, H. W. (1997). Experimental evidence of characteristic relations of type-I intermittency in an electronic-circuit. *Physical Review E*, 56:2573–2577.

Kinsler, L. E., Frey, A. R., Coppens, A. B., and Sanders, J. V. (1982). *Fundamentals of Acoustics*. Wiley.

Kruel, Th.-M., Eisworth, M., and Schreider, F. W. (1993). Computation of Lyapunov spectra: Effect of interactive noise and application to a chemical oscillator. *Physica D*, 63:117–137.

Krylov, N. and Bogoliubov, N. (1949). *Introduction to Non-Linear Mechanics*. Princeton University Press.

Lathrop, D. E. and Kostelich, E. J. (1989). Characterization of an experimental strange attractor by periodic orbits. *Physical Review A*, 40:4028–4031.

Levitan, E. S. (1960). Forced oscillation of a spring-mass system having combined Coulomb and viscous damping. *Journal of the Acoustical Society of America*, 32:1265–1269.

Li, T. Y. and Yorke, J. A. (1975). Period three implies chaos. *American Mathematical Monthly*, 82:985–992.

Liebert, W. and Schuster, H. G. (1989). Proper choice of the time delay for the analysis of chaotic time series. *Physics Letters A*, 142:107–111.

Linsay, P. S. (1981). Period-doubling and chaotic behavior in a driven anharmonic oscillator. *Physical Review Letters*, 47:1349–1352.

Lorenz, E. N. (1963). Deterministic nonperiodic flow. *Journal of the Atmospheric Sciences*, 20:130–141.

Lorenz, E. N. (1984). The local structure of a chaotic attractor in four dimensions. *Physica D*, 13:90–104.

Luenberger, D. G. (1979). *Introduction to Dynamic Systems: Theory, Models and Applications*. Wiley.

Lurie, H. (1952). Lateral vibrations as related to structural stability. *Journal of Applied Mechanics*, 19:195–204.

Lyapunov, A. M. (1947). *The General Problem of the Stability of Motion*. Princeton University Press.

MacRobie, F. A. and Thompson, J. M. T. (1991). Lobe dynamics and the escape from a potential well. *Proceedings of the Royal Society of London A*, 435:659–672.

Mandelbrot, B. B. (1983). *The Fractal Geometry of Nature*. W. H. Freeman.

Marek, M. and Schreider, I. (1991). *Chaotic Behavior of Deterministic Dissipative Systems*. Cambridge University Press.

Marion, J. B. and Thornton, S. T. (1988). *Classical Dynamics of Particles and Systems*. Harcourt Brace Jovanovich.

Math/Library, IMSL. (1989). IMSL problem solving software systems. Technical report, Houston.

Matlab (1989). Users guide. Technical report, The Math Works.

Matsumoto, T. (1987). Chaos in elecronic circuits. *Proceedings of the IEEE*, 75:1033–1057.

Matsumoto, T., Chua, L. O., and Tanaka, S. (1984). Simplest chaotic non-autonomous circuit. *Physical Review A*, 30:1155–1157.

May, R. M. (1976). Simple mathematical models with very complicated dynamics. *Nature*, 261:459–467.

McLachlan, N. W. (1964). *Theory and Applications of Mathieu Functions*. Dover.

Mees, A. (1981). *Dynamics of Feedback Systems*. Wiley.

Meirovitch, L. (1997). *Principles and Techniques of Vibrations*. Prentice- Hall.

Melnikov, V. K. (1963). On the stability of the center for time periodic solutions. *Transactions of the Moscow Mathematics Society*, 12:1–57.

Mitropolskii, Y. A. (1965). *Problems of the Asymptotic Theory of Nonstationary Vibrations*. Daniel Davey and Co.

Moon, F. C. (1985). Fractal basin boundaries and homoclinic orbits for periodic motion in a two-well potential. *Physical Review Letters*, 55:1439–1442.

Moon, F. C. (1992). *Chaotic and Fractal Dynamics, An Introduction for Applied Scientists and Engineers*. Wiley.

Moon, F. C. and Holmes, P. J. (1979). A magneto-elastic strange attractor. *Journal of Sound and Vibration*, 65:275–296.

Moon, F. C. and Holmes, W. T. (1985). Double Poincaré sections of a quasi-periodically forced, chaotic attractor. *Physics Letters*, 111:157–160.

Mullin, T. (1993). *The Nature of Chaos*. Oxford University Press.

Murphy, K. D. (1994). *Theoretical and experimental studies in nonlinear dynamics and stability of elastic structures*. PhD thesis, Duke University.

Murphy, K. D., Bayly, P. V., Virgin, L. N., and Gottwald, J. A. (1994). Measuring the stability of periodic attractors using perturbation induced transients: application to two nonlinear oscillators. *Journal of Sound and Vibration*, 172:85–102.

Murphy, K. D., Virgin, L. N., and Rizzi, S. A. (1996a). Characterizing the dynamic response of a thermally loaded, acoustically excited plate. *Journal of Sound and Vibration*, 196:635–658.

Murphy, K. D., Virgin, L. N., and Rizzi, S. A. (1996b). Experimental snap-through boundaries for acoustically excited, thermally buckled plates. *Experimental Mechanics*, 36:312–317.

Murphy, K. D., Virgin, L. N., and Rizzi, S. A. (1997). The effect of thermal prestress on the free vibration characteristics of clamped rectangular plates: Theory and experiment. *Journal of Vibration and Acoustics*, 119:243–249.

Nayfeh, A. H. and Balachandran, B. (1995). *Applied Nonlinear Dynamics*. Wiley.

Nayfeh, A. H. and Mook, D. T. (1978). *Nonlinear Oscillations*. Wiley.

Nerenberg, M. A. and Essex, C. H. (1990). Correlation dimension and systematic geometric effects. *Physical Review A*, 42:7065–7074.

Newland, D. A. (1984). *An Introduction to Random Vibrations and Spectral Analysis*. Longman.

Ng, C. F. and White, R. G. (1988). Dynamic behavior of postbuckled isotropic plates under in-plane compression. *Journal of Sound and Vibration*, 120:1–18.

Nichols, J. M. (1999). Classification of experimental dynamical systems through time series analysis. Master's thesis, Duke University.

Nordmark, A. B. (1991). Non-periodic motion caused by grazing incidence in an impact oscillator. *Journal of Sound and Vibration*, 145:279–297.

Nordmark, A. B. (1992). Effects due to low velocity impact in mechanical oscillators. *International Journal of Bifurcations and Chaos*, 2:597–605.

Nusse, H. E., Ott, E., and Yorke, J. A. (1995). Saddle-node bifurcations on fractal basin boundaries. *Physical Review Letters*, 75:2482–2485.

Nusse, H. E. and Yorke, J. A. (1994). *Dynamics: Numerical Explorations*. Springer-Verlag.

Nusse, H. E. and Yorke, J. A. (1996). Wada basins and basin cells. *Physica D*, 90:242–261.

Ogata, K. (1998). *System Dynamics*. Prentice-Hall.

Oppenheim, A. V. and Schafer, R. W. (1975). *Digital Signal Processing*. Prentice-Hall.

Ott, E. (1993). *Chaos in Dynamical Systems*. Cambridge University Press.

Ott, E., Grebogi, C., and Yorke, J. A. (1990). Controlling chaos. *Physical Review Letters*, 64:1196–1199.

Parker, T. S. and Chua, L. O. (1989). *Practical Numerical Algorithms for Chaotic Systems*. Springer-Verlag.

Parlitz, U. and Lauterborn, W. (1985). Superstructure in the bifurcation set of the Duffing equation $\ddot{x} + d\dot{x} + x + x^3 = f \cos(\omega t)$. *Physics Letters*, 107A:351–355.

Peterka, F. and Vacik, J. (1992). Transition to chaotic motion in mechanical systems with impacts. *Journal of Sound and Vibration*, 154:95–115.

Pezeshki, C. and Dowell, E. H. (1987). An examination of initial condition maps for the sinusoidally excited buckled beam modeled by Duffing's equation. *Journal of Sound and Vibration*, 117:291–232.

Pippard, A. B. (1985). *Response and Stability*. Cambridge University Press.

Plaut, R. H., Gentry, J. J., and Mook, D. T. (1990). Resonances in the non-linear structural vibrations involving two external excitations. *Journal of Sound and Vibration*, 140:371–379.

Plaut, R. H., HaQuang, N., and Mook, D. T. (1986). Simultaneous resonances in the non-linear structural vibrations under two-frequency excitation. *Journal of Sound and Vibration*, 106:361–376.

Poincaré, H. (1921). Analyse des travaux scientifiques de Henri Poincaré faites par lui-même. *Acta Mathematica*, 38:1–135.

Poston, T. and Stewart, I. (1978). *Catastrophe Theory and Its Applications*. Pitman.

Press, W. H., Flannery, B. P., Teukolsky, S. A., and Vetterling, W. T. (1992). *Numerical Recipes*. Cambridge University Press.

Rabinowicz, E. (1959). The intrinsic variables affecting the stick-slip process. *Proceedings of the Royal Society of London A*, 71:668–675.

Raman, A. P., Bajaj, A. K, and Davies, P. (1996). On the slow transition across bifurcations in some classical nonlinear systems. *Journal of Sound and Vibration*, 192:835–865.

Raty, R., von Boehm, J., and Isomaki, H. M. (1986). Chaotic motion of a periodically driven particle in an asymmetric potential well. *Physical Review A*, 34:4310–4315.

Rayleigh, J. W. S. (1945). *The Theory of Sound*. Dover.

Rega, G., Benedettini, F., and Salvatori, A. (1991). Periodic and chaotic motions of an unsymmetrical oscillator in nonlinear structural dynamics. *Chaos, Solitons and Fractals*, 1:39–54.

Reynolds, O. (1883). On the experimental investigation of the circumstances which determine whether the motion of water be direct or sinuous, and the law of resistance in parallel channels. *Philosophical Transactions of the Royal Society of London*, 174:935–982.

Robertson, J. (1996). *Engineering Mathematics with Maple*. McGraw-Hill.

Robinson, F. N. H. (1989). Experimental observation of the large-amplitude solutions of Duffing's and related equations. *IMA Journal of Applied Mathematics*, 42:177–201.

Romeiras, F. J. and Ott, E. (1987). Strange nonchaotic attractors of the damped pendulum with quasi-periodic forcing. *Physical Review A*, 35:4404–4413.

Roy, R., Murphy, T. W., Maier, T. D., and Gills, Z. (1992). Dynamical control of a chaotic laser: Experimental stabilization of a globally coupled system. *Physical Review Letters*, 68:1259–1262.

Ruelle, D. and Takens, F. (1971). On the nature of turbulence. *Cummunications in Mathematical Physics*, 20:167–192.

Schmidt, G. (1986). Onset of chaos and global analytical solutions for Duffing's oscillator. *Zeitschrift fur Angewandte Mathematik und Mechanik*, 66:129–140.

Schmidt, G. and Tondl, A. (1986). *Non-Linear Vibrations*. Cambridge University Press.

Seydel, R. (1991). Tutorial on continuation. *International Journal of Bifurcation and Chaos*, 1:3–11.

Seydel, R. (1994). *Practical Bifurcation and Stability Analysis*. Springer.

Shahian, B. and Hassul, M. (1993). *Control System Design Using Matlab*. Prentice-Hall.

Shaw, S. W. and Haddow, A. G. (1992). On 'roller-coaster' experiments for nonlinear oscillators. *Nonlinear Dynamics*, 3:375–384.

Shaw, S. W. and Holmes, P. J. (1983). A periodically forced piecewise linear oscillator. *Journal of Sound and Vibration*, 90:129–155.

Shenker, S. J. (1982). Scaling behavior in a map of a circle onto itself. *Physica D*, 5:405–411.

Slade, K. N., Virgin, L. N., and Bayly, P. V. (1997). Extracting information from interimpact intervals in a mechanical oscillator. *Physical Review E*, 56:3705–3708.

Smith, P. and Smith, R. C. (1990). *Mechanics*. Wiley.

Soliman, M. S. and Thompson, J. M. T. (1989). Integrity measures quantifying the erosion of smooth and fractal basins of attraction. *Journal of Sound and Vibration*, 135:453–475.

Soliman, M. S. and Thompson, J. M. T. (1991). Basin organization prior to a tangled saddle-node bifurcation. *International Journal of Bifurcation and Chaos*, 1:107–118.

Soliman, M. and Thompson, J. M. T. (1992). Global dynamics underlying sharp basin erosion in nonlinear driven oscillators. *Physical Review A*, 45:3425–3431.

Steidel, R. F. (1989). *Introduction to Mechanical Vibrations*. Wiley.

Stengel, R. F. (1986). *Stochastic Optimal Control*. Wiley Interscience.

Stewart, H. B. and Ueda, Y. (1991). Catastrophes with indeterminate outcome. *Proceedings of the Royal Society of London A*, 432:113–123.

Stewart, I. (1989). *Does God Play Dice?* Blackwell.

Stoker, J. J. (1992). *Nonlinear Vibrations in Mechanical and Electrical Systems*. Wiley Classics Library. Originally published in 1950.

Strogatz, S. H. (1994). *Nonlinear Dynamics and Chaos*. Addison-Wesley.

Swinney, H. L. (1983). Observation of order and chaos in nonlinear systems. *Physica D*, 7:3–15.

Swinney, H. L. and Gollub, J. P. (1978). Transition to turbulence. *Physics Today*, 31:41–49.

Szemplinska-Stupnicka, W. (1988). Bifurcations of harmonic solutions leading to chaotic motion in the softening type Duffing's oscillator. *International Journal of Non-Linear Mechanics*, 23:257–277.

Tabor, D. (1981). Friction: The present state of our understanding. *Journal of Lubrication Technology*, 103:169–179.

Tauchert, T. R. (1991). Thermally induced flexure, buckling, and vibration of plates. *Applied Mechanics Reviews*, 44:347–360.

Tel, T. (1990). Transient chaos. In *Directions in Chaos*. World Scientific.

Theiler, J. (1986). Spurious dimension from correlation algorithms applied to limited time-series data. *Physical Review A*, 34:2427–2432.

Theiler, J. (1990). Estimating fractal dimension. *Journal of the Optical Society of America*, 7:1055–1073.

Thompson, J. M. T. (1982). *Instabilities and Catastrophes in Science and Engineering*. Wiley.

Thompson, J. M. T. (1989). Chaotic phenomena triggering the escape from a potential well. *Proceedings of the Royal Society of London A*, 421:195–225.

Thompson, J. M. T. (1992). Global unpredictability in nonlinear dynamics: Capture, dispersal, and the indeterminate bifurcations. *Physica D*, 58:260–272.

Thompson, J. M. T. and Ghaffari, R. (1983). Chaotic dynamics of an impact oscillator. *Physical Review A*, 27:1741–1743.

Thompson, J. M. T. and Hunt, G. W. (1984). *Elastic Instability Phenomena*. Wiley.

Thompson, J. M. T. and Soliman, M. S. (1990). Fractal control boundaries of driven oscillators and their relevance to safe engineering design. *Proceedings of the Royal Society of London A*, 428:1–13.

Thompson, J. M. T. and Stewart, H. B. (1986). *Nonlinear Dynamics and Chaos*. Wiley.

Thompson, J. M. T. and Virgin, L. N. (1986). Predicting a jump to resonance using transient maps and beats. *International Journal of Non-Linear Mechanics*, 21:205–216.

Thomson, W. T. (1981). *Theory of Vibration with Applications*. Prentice-Hall.

Todd, M. D. (1996). *Local and global dynamical behavior in nonlinear mechanical models: Theory and experiments*. PhD thesis, Duke University.

Todd, M. D. and Virgin, L. N. (1997a). An experimental impact oscillator. *Chaos, Solitons and Fractals*, 8:699–714.

Todd, M. D. and Virgin, L. N. (1997b). An experimental verification of basin metamorphoses in a nonlinear mechanical system. *International Journal of Bifurcation and Chaos*, 7:1337–1357.

Todd, M. D., Virgin, L. N., and Gottwald, J. A. (1996). The nonstationary transition through resonance. *Nonlinear Dynamics*, 10:31–48.

Tongue, B. H. (1987). On obtaining global nonlinear system characteristics through interpolated cell mapping. *Physica D*, 28:401–408.

Tongue, B. H. (1996). *Principles of Vibration*. Oxford University Press.

Trickey, S. T. (1997). *Stabilizing chaotic systems: Towards applications*. Master's thesis, Duke University.

Trickey, S. T. and Virgin, L. N. (1998). Bottlenecking phenomena near a saddle-node remnant in a Duffing oscillator. *Physics Letters A*, 248:185–190.

Troger, H. and Steindl, A. (1991). *Nonlinear Stability and Bifurcation Theory: An Introduction for Engineers and Applied Scientists*. Springer-Verlag.

Tseng, W. Y. and Dugundji, J. (1971). Nonlinear vibrations of a buckled beam under harmonic excitation. *Journal of Applied Mechanics*, 38:467–476.

Tufillaro, N. B., Abbot, T., and Reilly, J. (1992). *An Experimental Approach to Nonlinear Dynamics and Chaos*. Addison-Wesley.

Ueda, Y. (1980). Steady motions exhibited by Duffing's equation: A picture book of regular and chaotic motions. In Holmes, P. J., ed., *New Approaches to Nonlinear Problems in Dynamics*. SIAM.

Ueda, Y., Yoshida, S., Stewart, H. B., and Thompson, J. M. T. (1990). Basin explosions and escape phenomena in the twin-well Duffing oscillator: compound global bifurcations organizing behaviour. *Philosophical Transactions of the Royal Society of London*, 332:169–186.

van der Pol, B. (1934). The nonlinear theory of electric oscillations. *Proceedings of the Institute of Radio Engineers*, 22:1051–1086.

van Dooren, R. (1988). On the transition from regular to chaotic behaviour in the Duffing oscillator. *Journal of Sound and Vibration*, 123:327–339.

Virgin, L. N. (1985). The dynamics of symmetric postbuckling. *International Journal of Mechanical Science*, 27:235–248.

Virgin, L. N. (1986). Parametric studies of the dynamic evolution through a fold. *Journal of Sound and Vibration*, 147:99–109.

Virgin, L. N. (1987). The nonlinear rolling response of a vessel including chaotic motions leading to capsize in regular seas. *Applied Ocean Research*, 9:89–95.

Virgin, L. N. (1988). Approximate criterion for capsize based on deterministic dynamics. *Dynamics and Stability of Systems*, 4:55–70.

Virgin, L. N., Dowell, E. H., and Conner, M. D. (1999). On the evolution of deterministic non-periodic behavior of an airfoil. *International Journal of Nonlinear Mechanics*, 34:499–514.

Virgin, L. N., Fielder, W. T., and Plaut, R. H. (1996). Transient motion and overturning of a rocking block on a seesawing foundation. *Journal of Sound and Vibration*, 191:177–187.

Virgin, L. N. and Murphy, K. D. (1994). On the behavior of characteristic multipliers through a period-doubling sequence. *Journal of Sound and Vibration*, 169:699–703.

Virgin, L. N., Plaut, R. H., and Cheng, C.-C. (1992). Prediction of escape from a potential well under harmonic excitation. *International Journal of Nonlinear Mechanics*, 27:357–365.

Virgin, L. N., Todd, M. D., Begley, C. J., Trickey, S. T., and Dowell, E. H. (1998). Basins of attraction in experimental nonlinear oscillators. *International Journal of Bifurcation and Chaos*, 8:521–533.

White, R. G. (1971). Evaluation of the dynamic characteristics of structures by transient testing. *Journal of Sound and Vibration*, 15:147–161.

Wiesenfeld, K. (1985). Noisy precursors of nonlinear instabilities. *Journal of Statistical Physics*, 38:1071–1097.

Wiggins, S. (1987). Chaos in the quasiperiodically forced Duffing oscillator. *Physics Letters A*, 124:138–142.

Wiggins, S. (1990). *An Introduction to Applied Dynamical Systems Theory and Chaos*. Springer-Verlag.

Wolf, A., Swift, J. B., Swinney, H. L., and Vastano, J. A. (1985). Determining Lyapunov exponents from a time series. *Physica D*, 16:285–317.

Wolfram, S. (1996). *The Mathematica Book*. Cambridge University Press.

Worden, K. (1996). On jump frequencies in the response of the Duffing oscillator. *Journal of Sound and Vibration*, 198:522–525.

Yasuda, K. and Kamiya, K. (1990). Identification of a nonlinear beam (proposition of an identification technique). *Japanese Society of Mechanical Engineering (Series III)*, 33:535–540.

Yun, C.-B. and Shinozuka, M. (1980). Identification of nonlinear structural dynamics systems. *Journal of Structural Mechanics*, 8:187–203.

Index